Computer Science Foundations and Applied Logic

Computer Science Foundations and Applied Logic is a growing series that focuses on the foundations of computing and their interaction with applied logic, including how science overall is driven by this. Thus, applications of computer science to mathematical logic, as well as applications of mathematical logic to computer science, will yield many topics of interest. Among other areas, it will cover combinations of logical reasoning and machine learning as applied to AI, mathematics, physics, as well as other areas of science and engineering. The series (previously known as *Progress in Computer Science and Applied Logic, https://www.springer.com/series/4814*) welcomes proposals for research monographs, textbooks and polished lectures, and professional text/references. The scientific content will be of strong interest to both computer scientists and logicians.

Mariëlle Stoelinga · Enno Ruijters ·
Pavel Krčál

Concise Guide to Fault Tree Analysis

Models, Methods and Algorithms

 Birkhäuser

Mariëlle Stoelinga 🆔
Department of Computer Science
University of Twente
Enschede, The Netherlands

Enno Ruijters
BetterBe B.V.
Enschede, The Netherlands

Pavel Krčál
RiskSpectrum AB
Stockholm, Sweden

ISSN 2731-5754   ISSN 2731-5762   (electronic)
Computer Science Foundations and Applied Logic
ISBN 978-3-031-78286-2   ISBN 978-3-031-78287-9   (eBook)
https://doi.org/10.1007/978-3-031-78287-9

This book is published under the imprint Birkhäuser, www.birkhauser-science.com by the registered
company Springer Nature Switzerland AG
The registered company address is: Gewerbestrasse 11, 6330 Cham, Switzerland

If disposing of this product, please recycle the paper.

All models are wrong,
but some are useful.
G. E. P. Box

Foreword

Fault Tree Analysis (FTA) is a failure analysis method used by Dependability and Safety Engineers in multiple industrial domains and particularly in Aerospace. In the European Space Agency (ESA), FTAs are mostly used in Human Spaceflight applications, as inputs to Probabilistic Risk Assessment (PRA) for the quantification of Loss of Mission and/or Loss of Crew (LoM/LoC). This technique is also used in other space projects in which Feared Events need to be analysed (e.g. navigation system). Recently, due to the ESA zero debris policy and to contribute to the sustainable space, FTAs are now also being used for assessing combination of failures that may lead to in-orbit collisions.

A Fault Tree is a deductive and top-down technique that allows to analyse combinations of failures leading to the systems' top event or undesired state. The analysis is both qualitative analysis (i.e., failure tolerance justification) and quantitative, with the computation of the probability of occurrence of the Feared Events and supporting the probabilistic apportionment to the different contributors.

The use of FTA offers several advantages, which makes this method very popular within the Dependability and Safety community. The main advantage is the step-by-step approach which facilitates a concurrent analysis together with the design team and a validation of the modelling. FTA is also very powerful visually because of its logical, systematic and comprehensive approach making it appealing and easy to understand for (non-expert) decision makers.

Reading this guide, you will discover the many benefits of FTA and the qualitative and quantitative analyses associated with it. It offers a great journey into the world of FTA: the presentation is crystal clear, comprehensive and well-underpinned, with numerous examples illustrating the various concepts and techniques. Through this guide, you will be able to learn the multifaceted techniques of FTAs and how to apply them in your application/project.

Fabrice Cosson
Head of the RAMS Section
European Space Agency (ESA)
Noordwijk, The Netherlands

Preface

This book is intended for anybody interested in risk analysis.

Risk management is an important activity to prevent failures and accidents. Actually, we do risk management on a daily basis: we send our cars to the garage for check-ups and maintenance; when going on vacation, we may purchase travel insurance to cover the cost of a medical treatment. At industrial scale, risk management happens at large. The consequences of failures, such as the leakage of a chemical substance or downtime in critical infrastructures, are higher and so are the costs of measures to prevent or repair them. For example, the Golden Gate Bridge is painted to prevent the high salt in the air to cause the steel to corrode or rust, costing $150,000 annually on paint only [1]. Thus, decision making in risk analysis is important. However, it is also complex: Golden gate bridge uses inspectors to identify bad sections, which are treated immediately. Modern engineering systems often consist of thousands of components. Understanding and managing their risks is far from trivial, since risks can have multiple sources, and arise due to a chain of effects and causes. Understanding these cause-effect relations and intervening at the right place is not easy.

To support decision making, a number of models have been put forward. These include risk registers, failure mode and effect analysis (FMEA), fishbone diagrams, as well as more engineering-oriented approaches such as reliability block diagrams, AADL and SysML. Each of these approaches has its advantages and disadvantages. Fault tree analysis stands out: It is widely used in practice, supports both quantitative and qualitative analysis, it breaks down system level failures into their constituent components, allowing to thoroughly understand root causes.

Like any risk analysis method, fault tree analysis must be performed with a suitable and clear purpose in mind, at the right moment in time (enabling measures to be implemented), and by the right people. If done properly, fault tree analysis can save lives.

Enschede, The Netherlands
Enschede, The Netherlands
Stockholm, Sweden

Mariëlle Stoelinga
Enno Ruijters
Pavel Krčál

Acknowledgments Several people have provided valuable and practical suggestions to improve earlier versions of this book: Ola Bäckström, Yanni Dong, Jaap van Ekris, Lisandro Jimenez Roa, Gea Kolk, Milan Lopuaä-Zwakenberg, Stefano Nicoletti, Reza Soltani.

Reference

1. Bjerklie S (2006) Rust never sleeps: keeping the golden gate bridge covered with paint is a major challenge for crews, management, and coatings systems. Met Finish 104(2):33–36, 10. 1016/S0026-0576(06)80008-5

Contents

Introduction

Reliability and risk engineering. Our society crucially depends on complex engineering systems such as (nuclear) power plants, airplanes, data centers, robots, smart grids, self-driving cars, and many more. To ensure that such systems operate safely and reliably, proper risk assessment is a crucial activity. The key is to identify the highest risks and devise effective interventions to mitigate these. This holds true during all phases of the system's life cycle: design, implementation, operation, and dismantling. Moreover, the increased constraints imposed by international standards, together with the ever-growing penetration of AI, make rigorous and powerful risk assessment more important than ever.

Over the years, various techniques have been developed to analyze system safety and reliability. One of the most relevant techniques is fault tree analysis. Every day, millions of engineers are using fault trees to model their systems and analyze the probability of failures. Fault tree analysis is applied to many safety-critical systems and their use is often mandated, for instance, by the Federal Aviation Administration (FAA), the Nuclear Regulatory Commission (NRC), and for software development in aerospace systems.

Fault tree analysis is part of the broader fields of RAMS (reliability, availability, maintenance, and safety) and reliability engineering, and it is often applied in probabilistic safety assessment. Given its importance in these fields, various ISO and IEC standards recommend fault tree analysis. Examples are the standard IEC 61508 on functional safety, ISO 14971 for medical devices, and ISO 26262 for autonomous driving. Fault tree analysis itself has been standardized by the International Electronic Commission as the standard IEC 61025.

Fault trees. Fault tree analysis is a graphical technique to identify potential causes of system failures by systematically breaking down high-level system failures into their causal factors. As such, fault trees help to understand, document, analyze, and communicate potential system failures and their impact. Thus, they are a widely

M. Stoelinga et al., *Concise Guide to Fault Tree Analysis*, Computer Science Foundations and Applied Logic, https://doi.org/10.1007/978-3-031-78287-9_1

used tool in decision making, by answering questions such as: is the system safe and reliable enough? What are the most vulnerable areas? What are actions to reduce system risk? How do various design alternatives compare?

Thus, the goal of fault tree analysis can be phrased as follows.

> *The main purpose of fault tree analysis is to make risk-informed decisions regarding system design and operations.*

Typical design decisions concern the overall system architecture, the level of redundancy and the required quality of components. Operational decisions concern, for example, the inspection frequencies, spare parts management, repairs, and replacements. This purpose can include other purposes, such as certification, documentation, and diagnosis.

Analysis techniques. Fault trees support their purpose in two ways. Fault tree diagrams provide a graphical means to understand a system's failure behavior and the relation between the various failure causes. Various analysis techniques provide additional insights in these failure causes and their impact. Fault tree analysis can be either qualitative or quantitative.

- *Qualitative techniques* point to critical paths and failure causes. Typical techniques are the identification of so-called minimal cut sets and common cause factors.
- *Quantitative techniques* aim to compute so-called dependability metrics. These are key performance indicators that quantify how well a system performs in terms of dependability. Numerous analytical and statistical methods are available to support their computation. Common metrics include the *system unreliability*, i.e., the probability that a system fails within a its mission time; the *system availability*, i.e., its average up-time; the *mean time to failure (MTTF)*, i.e., the average time for a system to fail. Furthermore, *importance factors* quantify the impact of a particular component on the system dependability.

Various software tools exist to support reliability engineers and risk analysis in the modeling and analysis of fault trees. A key issue is the sheer size of fault tree models that appear in certain applications. In fact, fault trees of all sizes occur in practice. Fault trees in some applications feature a few dozens or a few hundreds of component failures, while others handle more than 50,000 components. In the latter case, efficient algorithms are of crucial importance. Yet, exact analysis is not always possible, so for large trees exact results are approximated.

This book

A wide range of different fault tree variants exist, featuring many different modeling constructs. In this book, we focus on the basic concepts and principles, setting the scope as follows:

> *This book covers static fault trees.*

Static fault trees appeal as a powerful, yet relatively simple formalism. These express causal dependencies between failure causes via Boolean gates, such as AND and OR. These gates are expressive enough to model many dependability patterns, such as redundancy and failure conditions. At the same time, they are simple enough to enable powerful analysis algorithms that can handle large trees: most of them via exact calculations, and the largest trees by approximation.

Moreover, static fault trees can be studied in many different settings: with or without probabilistic failure distributions, with failures over time or at a single instance, with and without repairs, etc. yielding a wide plethora of analysis possibilities. Static fault trees have been extended in several ways, catering for more complex dependability patterns. The last chapter of this book covers several of these extensions.

Objectives. This book serves as a practical resource for anyone looking to master the fundamentals and advanced aspects of fault tree analysis. The book is designed to provide you with a comprehensive understanding of the following key objectives:

- *Understand what fault trees are and how they work:* Gain a clear understanding of fault trees, their structure, and their role in system reliability and safety analysis.
- *Practices of creating fault trees and performing and organizing a fault tree analysis:* Learn step-by-step methods for constructing fault trees and organizing an effective fault tree analysis.
- *Understand qualitative analysis and its interpretation:* Explore techniques for qualitative analysis within fault tree analysis, including how to interpret and apply the results.
- *Understand quantitative analysis and its interpretation:* Delve into quantitative analysis methods, learning how to calculate probabilities and assess risks.
- *Specifics around Common Cause Failures (CCFs):* Understand the impact of CCFs on system reliability and how to address them within fault trees analysis.

How to read this book?

Overview of chapters. The main purpose of this book is to provide a comprehensive overview of the principles and techniques for fault tree analysis. To achieve this goal, this book guides the reader through the following questions.

Part I: *What are fault trees and how are they used to analyze and enhance system dependability?*

- Chapter 2, *Fault trees at a glance* presents a bird's eye view on fault trees analysis, describing core modeling and analysis concepts, elaborated in subsequent chapters.
- Chapter 3, *Algorithmic building blocks* introduces several foundational techniques.

Part II: *How to perform a fault tree analysis?*

This part describes the process of performing an effective fault tree analysis.

- Chapter 4, *The role of fault tree analysis in decision making* explains the various purposes a fault tree analysis can have.
- Chapter 5, *The process of fault tree analysis* presents a structured procedure to set up the whole analysis process, following the ISO 31000 and IEC 61025 standards.
- Chapter 6, *Creating fault tree models* describes good practices to create relevant and faithful fault trees.

Part III: *How to identify the most impactful system vulnerabilities?*

Cut set analysis is the one of the most fundamental techniques in fault tree analysis, pinpointing root causes and critical system parts. Three chapters are devoted to this topic.

- Chapter 7, *Minimal cut sets* presents the concept of cut sets and several cut set metrics to rank their importance for the system dependability.
- Chapter 8, *Quantification of minimal cut set lists* delves into methods for probabilistic fault tree analysis by computing cut set probabilities.
- Chapter 9, *Computing minimal cut sets* focuses on methods for computing a fault tree's cuts sets.

Part IV: *Which methods exist to quantify the dependability of a system?*

This part is concerned with probabilistic fault tree analysis. Common dependability metrics are discussed, such as the system reliability, availability, mean time to failure, mean time to repair.

- Chapter 10, on *Failure probabilities* studies a fault tree's probabilistic behavior in a fixed time frame.
- Chapter 11, on *Fault trees without repairs* studies the evolution of the failure behavior over time, but with no repairs.
- Chapter 12, on *Fault tree with repairs* studies the evolution of the failure behavior over time, with repairs.
- Chapter 13, *Importance measures* discusses measures that indicate which system parts contribute most to the system's failure behavior.

Part V: *What variations and related models exist?*

Numerous variations and extensions for fault trees exist. We discuss three.

- Normally, fault trees assume that all component failures are independent. Chapter 14, *Common cause failure* discuss the case where this assumption does not hold.
- Chapter 15, *Non-coherent fault trees* discusses fault tree with negated gates. These require different analysis and interpretation, especially for their cut sets.
- Chapter 16, *Beyond static fault trees* discusses a variety of extensions.

What to read?

We aimed to create a book that is both accessible to practitioners and valuable to developers of fault tree analysis methods. We accomplished this by clearly describing fault tree analysis concepts and methods, offering practical interpretations, and incorporating numerous examples throughout the book.

An advantage of fault tree analysis is its mathematical precision, ensuring accurate and unambiguous analysis results. This book embraces the mathematical approach and provides formulas for all concepts and techniques. Textual descriptions accompany all mathematical notations, allowing readers to skip the math if they prefer.

Readers interested in: Applying fault tree analysis

People with an interest in performing fault tree analysis are invited to read

- Part I, Chap. 2 serves as a basis for the entire book.
- Part II, Chaps. 4–6 cover the entire process of fault tree analysis.
- Part III covers the fundamental technique of cut set analysis. Its first chapter Chap. 7 presents the main concepts and techniques; subsequent chapters are more technical.
- Part IV on probabilistic fault tree analysis explains the mathematics behind probabilistic analysis. We suggest the first part of Chap. 10. Further, Chap. 13 is a stand-alone chapter that focuses on quantifying the importance of components.
- In Part V, Chaps. 14 and 15 are indispensable when handling common cause failures and non-coherency.

Throughout the book, we provide both textual explanations, examples, and the underlying mathematical formulations. **All mathematical notation can be skipped.**

Readers interested in: Understanding the techniques behind fault tree analysis

People with an interest in the background of the fault tree analysis techniques are invited to read

- Part I, Chap. 2 serves as a basis for the entire book.
- In Part II, we suggest the first Chap. 4 to understand how fault tree analysis is organized.
- Part III covers the fundamental technique of cut set analysis. Its first chapter Chap. 7 presents the main concepts and techniques; subsequent chapters provide a deep dive into the underlying computation and quantification methods.
- Part IV on probabilistic fault tree analysis covers the fundamentals behind several dependability metrics, under different regimes (point-probabilities, time dependent probability distribution, repairs or no repairs).
- In Part V, Chaps. 14 and 15 are indispensable when handling common cause failures and non-coherency.

Further selections can be made based on the topics of interest.

We wish all readers a pleasant journey.

Part I
Overview

Fault Tree Analysis at a Glance

2

Fault trees are hierarchical, logic-based diagrams that model how systems fail. They operate by breaking down complex failures into subcauses until the basic, atomic causes are identified. These atomic causes are commonly known as root causes; nonetheless, determining what constitutes a root cause is a matter of choice.

Example 1 (Running example: the stranding of a road trip)

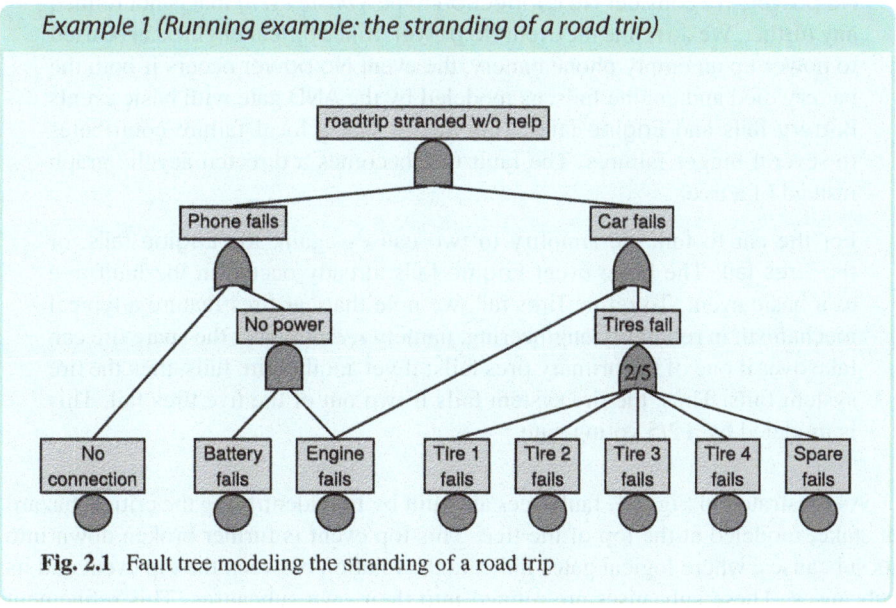

Fig. 2.1 Fault tree modeling the stranding of a road trip

M. Stoelinga et al., *Concise Guide to Fault Tree Analysis*, Computer Science Foundations and Applied Logic, https://doi.org/10.1007/978-3-031-78287-9_2

Imagine you are on a road trip, say in the Moroccan desert. The place is remote, the roads are rough; this is not a place to get stranded without help. As a preventive measure, you brought your phone, to call emergency services in case of problems. Also, your car has a spare tire, to replace a flat tire. Key questions concerning the (technical) risks are: *Is this set up reliable enough? If not, what additional measures make sense? What are the most vulnerable system parts? How many failures can we expect, if we organize 1.000 of such road trips per year? How much maintenance should we do?* These are all questions that can be analyzed by fault tree analysis.

Starting point in fault tree analysis is a fault tree diagram. Figure 2.1 models the safety for our car trip. It starts with the top event, expressing our main safety concern. We choose Road trip stranded if a car problem occurs, while your phone is unavailable to call for help. The top event is further refined into the subcauses Phone fails and Car fails. Since both events must occur, we use an AND-gate to refine the top event Road trip stranded to the intermediate events Phone fails and Car fails.

The next step is to refine these intermediate events. For simplicity, we list only two causes to refine the Phone fails event: Either there is No connection or No power. We consider No connection to be a *basic event* that is not refined any further. We do refine the event No power: Since the car engine can be used to power up an empty phone battery, the event No power occurs if both the battery died and engine fails, as modeled by the AND gate with basic events Battery fails and Engine fails. This means that a local failure contributes to several bigger failures. The fault tree becomes a directed acyclic graph instead of a tree.

For the car to fail, we simplify to two causes again: the Engine fails, or the Tires fail. The basic event Engine fails already occurs in the fault tree as a basic event. To refine Tires fail, we note that car tires feature a typical mechanism in reliability engineering, namely *redundancy:* the spare tire can take over if one of the primary tires fails; if yet another tire fails, then the tire system fails. Thus, the tire system fails if two out of the five tires fail. This is modeled by a 2/5 voting gate.

As illustrated in Fig. 2.1, fault trees are built by first identifying the critical hazard at stake, modeled at the top of the tree. This top event is further broken down into its subcauses, where logical gates indicate the relation between the top event and its subcauses. These subcauses are refined into their own subcauses. This refinement process continues until the basic causes are found. These basic causes are not explored or refined further, and appear as leaves at the bottom of the tree.

The main purpose of fault tree analysis is to understand how systems fail, and leverage these insights to make risk-based decisions regarding system design and operations. This involves efforts to eliminate, mitigate, and/or minimize the factors contributing to failures. Two types of analysis exist to support these decisions: *Qualitative fault tree analysis* provides important insights in the system's failure behavior, its root causes and pinpoints critical paths and components. *Quantitative analysis* in computing focuses on calculating numerical values for the top event and an explanation of this top value, for example by ranking of the top value contributors.

This chapter provides an overview of fault tree analysis methodology, covering fault tree diagrams, analysis techniques, their role in decision-making, and the underlying mathematical principles. Alternative risk analysis methods and a brief historical overview are discussed.

Running example. Example 1 shows a simple fault tree for a mission failure of a road trip. We use it throughout the book to illustrate different aspects of fault tree modeling and analysis.

2.1   Fault Tree Diagrams

Fault trees are hierarchical, logic-based diagrams that model how systems fail, breaking down complex failures into subcauses until their basic causes are found. A fault tree diagram consists of three main ingredients, shown in Fig. 2.2: *Events* represent potential failures and failure causes in the system under study, depicted by boxes in the fault tree diagram. *Directed edges* represent the direct relation between a failure and its subcauses. Finally, *logical gates* indicate the logical relationship between an event and its immediate subcauses. In this manner, fault trees visualize how lower-level events lead to higher-level events, ultimately culminating in a system failure.

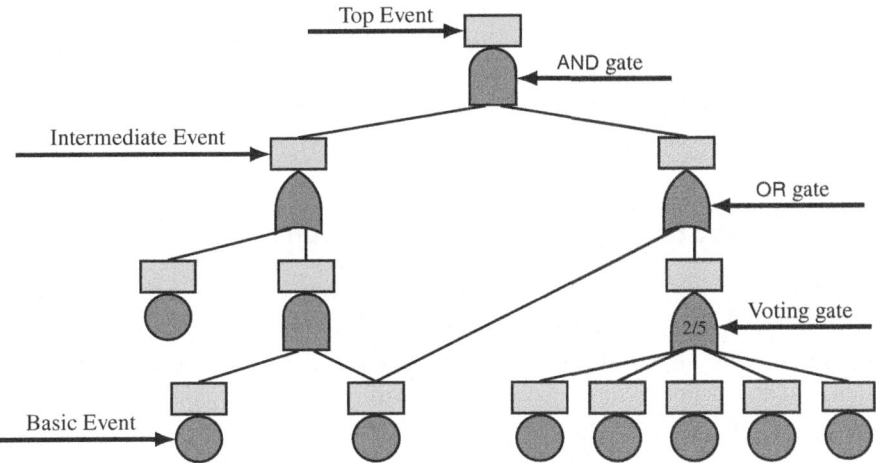

Fig. 2.2 Fault tree terminology

2.1.1   Events: Potential Failures and Conditions

Nodes in fault trees encode events, typically representing a possible failure of (a part of) the studied system. Events can be an atomic failure, such as a component failure mode, or a composite event representing a failure of a larger (sub-)system to provide the expected functionality. Besides actual failures, fault trees may include configuration or condition events as factors relevant for the system failure behavior, such as the temperature exceeding a certain threshold. Three types of events are distinguished: the top event, intermediate events, and basic events.

Top event. As a top-down methodology, fault trees are created by first identifying the critical failure of interest. Examples include a short-circuit fault in an electric power system, a stranded mission in a military operation, or a phone app not responding to user input. This event is placed at the top of the tree, and is therefore called the *top level undesired event*, or simply top level event, top undesired event, or top event.

Intermediate events. The top event is broken down into one or more *intermediate events*, which indicate the requirements or subcauses for the top event to occur. Intermediate events are drawn as rectangles in fault tree diagrams. Intermediate events are further refined into their subcauses, until the basic causes are found. These are called basic events and are associated with circles in the leaves of the tree.

Basic events. *Basic events* represent atomic failures. Which events are considered atomic might vary, even for one and the same system, depending on the analysis' purpose and scope. Basic events can encompass a wide range of occurrences, including failures (such as software, hardware, and human errors), cyberattacks, external conditions, or specific plant configurations. Examples include a gas turbine failing to start, an operator neglecting to reduce pressure in a vessel, a rupture caused by a seismic event, deliberate obstruction on a rail track, winter conditions, or maintenance activities.

Often, the basic events are equipped with quantitative information, such as failure probabilities, rates, repair or testing times, that serve as input to the quantitative analysis process. These quantitative entities are further discussed in Part IV.

Finally, since failures are the most common events, we often say that event E fails, rather than event E occurs.

2.1.2   Directed Edges: Dependencies Between Events

If one event E is refined into other events, then we call the latter events *children*, or *inputs* of E, and we call E the parent. In diagrams, we draw an edge from each input to its parent. The edge is meant to be directed, leading from the input to the parent. Often, and also in our pictures, these edges are drawn as lines, rather than as arrows. It is assumed that the lower event is input to the higher one.

Shared events. A single event can contribute to the failure of several other events, meaning one event has multiple parents. These events are called *shared* or *repeated*

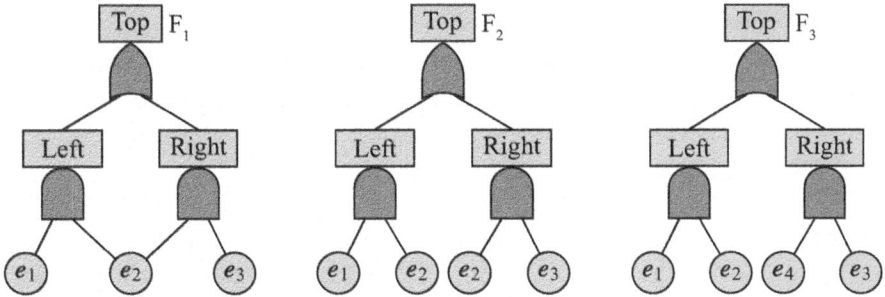

Fig. 2.3 Event e_2 is shared in fault trees F_1 and F_2. Fault tree F_3 has no shared events

events. In the road trip example, the basic event Engine fails is shared, as it contributes to both No power and Tires fail. Shared events can be represented in a fault tree either by repeating the event several times, or by linking it to all relevant parents, as illustrated in Fig. 2.3. In both cases, a single event has several parents. As detailed throughout the book, fault trees without repeated events are much easier to analyze. However, repeated events do occur often in practice.

2.1.3 Logic Gates: Causal Relationship Between Events

Fault tree gates establish the relationship between a top or an intermediate event and its direct causes. Specifically, gates specify the causal relation between occurrence of causes (children) and their directly related failure (parent). In other words, they describe how simpler failures propagate to a more complex failure. The relations treated in this book do not go beyond combinations of causes that can be described by Boolean logic. These are presented in Table 2.1.

Three main types of logic gates are used in fault trees: the AND-gate, OR-gate, and VOTING-gate.

AND-gates. The AND-gate is used when all causes must occur for the parent (intermediate or top) event to occur. In other words, the output event happens if all input events connected to the AND-gate happen. For example, if we have a system with two redundant pumps and one is sufficient to maintain the system function, then an AND-gate is used to connect these events to model the failure of the whole system.

OR-gates. The OR-gate is used if any one of the input events is sufficient to cause the output event. In other words, the output event happens if at least one of the input events connected to the OR-gate happens. If, for instance, a pump failure may result from either a mechanical or an electrical failure, then an OR-gate is used to connect these events.

VOTING-gates. The k/N voting gate requires that least k of the subcauses must be present for the undesired event to occur.

Table 2.1 Symbols for events and logical gates

Symbol	Name	Description
(circle)	Basic Event (BE)	Elementary failure that is not broken down into further subcauses
Label	Event name	Each basic, intermediate and top event is given a descriptive name that distinguishes them from other events. Basic events names may appear in their circle
(AND-gate)	AND-gate	Indicates that *all* subcauses must be present for the undesired event to occur
(OR-gate)	OR-gate	Indicates that *at least one* subcause must be present for the undesired event to occur
k/N	k/N-VOTING gate	Also called k-out-of-N gate, indicates that at least k of the N subcauses need to be present

Clearly, an OR-gate is the same as $1/N$-voting gate, and an AND-gate is the same as N/N-voting gate. Since OR-gates and AND-gates are much more common than voting gates, they have their own symbols.

Combinations of immediate causes may have a different effect on the functioning of the system depending on the context. For example, a system with four redundant subsystems might still succeed in providing the required function when at least one of the subsystems is working. A different system or even the same system in a different situation might require at least three subsystems to function, tolerating a single subsystem failure, but not a failure of two subsystems. For example, in normal circumstances, one pump may be enough to keep the water levels at acceptable heights, but in case of exceptional rain fall, several pumps may be required. Rules specifying the required functionality are called *success criteria* and are usually defined by the design engineers. Dependability analysts will use these to study propagation of failures through the system.

Additional gates. Numerous other gate types exist. Table 2.2 lists additional symbols for modeling convenience. Chapter 15 discusses Boolean gates expressing negation. Both are covered in this book. Chapter 16 examines various extensions requiring analysis techniques that fall outside the scope of this book.

Table 2.2 Additional symbols for modeling convenience

Symbol	Name	Description
	Transfer in	Indicates an input that is given by a different tree
	Transfer out	Indicates the output of a tree serves as input to another tree
	House event	A non-probabilistic event whose state is fixed before analysis
	Undeveloped event	An event that is not further refined due to lack of information
Condition	INHIBIT-gate	Indicates that a subcause only leads to the undesired event under a certain condition

2.1.4   Application Domains

Fault trees are a very generic and application-independent modeling technique, without specific constructs to model phenomena in particular domains. As such, they are deployed in many different industries, such as the nuclear power generation, rail, avionics and aerospace, automotive, water management. Clearly, creating faithful fault tree models does require domain-specific knowledge about the failure modes and mechanisms at hand. Several domain-specific guidelines exist to construct high-quality fault tree diagrams. These are discussed in Sect. 5.1.2.

2.2   Analysis of Fault Tree Diagrams

2.2.1   Purposes of Fault Tree Analysis

As stated in the introduction of this book, the main goal of fault tree analysis is to make risk-informed decisions regarding system design and operations. This overall purpose can be refined into several purposes. Chapter 5 details the most effective techniques for achieving each purpose.

- *Understanding and documentation.* By offering a systematic overview of system vulnerabilities and potential failure scenarios, fault trees enhance the understanding of failure behavior and accident scenarios and serve as valuable documentation for future reference.

- *Compliance and certification.* Safety-critical systems must often adhere to strict dependability requirements, formulated as dependability metrics, such as the reliability, availability, and mean time to failure (MTTF). Effective quantitative analysis methods to calculate these metrics make fault tree analysis a key tool for demonstrating compliance.
- *Design and operational decisions.* An important goal of risk analysis is to improve system dependability. Fault tree analysis supports this process by identifying and selecting cost-effective measures, i.e., to prevent or reduce the probability of the top event.
- *Diagnosis and monitoring.* After a system failure has occurred, fault trees can help understand how this failure happened, by systematically tracing back the failure causes in a fault tree diagram.

2.2.2 Fault Tree Analysis Techniques

While fault tree diagrams are essential for visualizing a system's potential failure behavior, a main strength of fault trees lies in their powerful analysis capabilities. These are based on solid mathematical foundations and comprise both qualitative and quantitative techniques.

Cut set analysis. Cut sets are a fundamental concept in fault tree analysis. They identify which combinations of failures, i.e., which basic events, make the top event to occur. In the road trip example, the set {Battery fails, Engine fails} is a cut set: Together Battery fails and Engine fails make the intermediate event No power occur, which in turn makes the event Phone fails happen. The basic event Engine fails also makes the event Car fails occur. Thus, both Phone fails and Car fails occur, making the top event Road trip stranded occur. Other cut sets are {No connection, Engine fails}, and {No connection, Tire2 fails, Tire3 fails}. Part III discusses cut sets in detail.

Qualitative techniques focus on combinations of basic failures, without quantifying probabilities of their occurrence. They typically explore questions like whether single failures can lead to a top event and if certain combinations of basic events occurs in majority of failure scenarios.

Cut sets provide essential qualitative information and enable better understanding of the system failure behavior by listing scenarios leading to a system failure. Cut sets can be also used to identify vulnerable components and rank dependability killers. A cut set with just one element indicates a single point of failure and deserves special attention. Cut sets of low order, i.e., with a few elements, also point to vulnerabilities, especially if the elements are likely to fail. One may argue that {No connection, Engine fails} is a vulnerability, since phone connections do fail, and then one only needs one additional failure to happen.

Quantitative techniques utilize numerical, statistical, data to reason about probabilities and importance of different risks and vulnerabilities. Basic events must be equipped with failure probabilities or rates and other reliability parameters, obtained from statistical methods or engineering judgment. The key quantitative question, essential for addressing other specific inquiries, concerns the probability of the top event occurring – known as the *top event probability*.

Cut sets serve again as a valuable tool, along with other methods such as Boolean Decision Diagrams. For example, ranking of cut sets according to their probability provides an overview of the most likely accident scenarios.

Other common dependability metrics include system availability, reliability, failure intensity, mean time to failure (MTTF), and mean time to repair (MTTR). These metrics quantify system risk levels. Furthermore, importance measures are used to reveal greatest risk contributors and importance of subsystems and components for accident mitigation. They can also analyze effectiveness of various measures to improve the system dependability. Part IV is devoted to dependability metrics.

Overall, fault trees provide a comprehensive approach to understanding and mitigating system failures, combining both qualitative and quantitative analyses to enhance system reliability and safety, improve risk understanding and facilitate risk-informed decisions. Various commercial and academic tools support both fault tree modeling and analysis. Section 5.3 covers several widely used options.

2.2.3   The Process of Fault Tree Analysis

Performing a fault tree analysis involves a number of steps: One starts with the preparation: setting the scope, analyzing the goals and team, and planning the analysis. Next, the top level undesired event is determined, as the overall focal point of the analysis. The fault tree is then constructed by refining the top event. A crucial step is collecting failure data required for quantitative analysis, followed by the fault analysis itself. This includes calculating cut sets, dependability metrics, and importance measures. After careful validation and review, these results guide the design and implementation of measures to eliminate, mitigate, or minimize the top level event.

Each of these steps is supported by guidelines and best practices, further elucidated in Chap. 5.1.

2.3   Mathematical Foundations

Mathematically, a fault tree is a directed acyclic graph G. The nodes (or vertices) in this graph represent the fault tree events. Its edges, written (v, v') or $v \rightarrow v'$, represent that event v is input to event v', i.e., v is a child of v'. In diagrams, v is depicted directly below v'. Each non-basic event is labeled by its gate type AND, OR, or $\text{VOT}(k/N)$.

Definition 1 A *fault tree* consists of the following components:

- A graph $G = (V, E)$ that represents the structure of the fault tree as a graph, where V is the set of nodes (or vertices) and E is the set of edges in G. We require that G is acyclic.
- A top level event *Top*. We require that *Top* is the unique root of G, i.e., the only node without outgoing edges.
- A function *Gate* : $V \setminus BE \rightarrow$ {AND, OR, VOT(k/N)} that assigns to each intermediate or top node its gate type.

Given a node, Inputs(v) denotes the set of all inputs (or children) of v', i.e., Inputs$(v') = \{v \in V \mid v \rightarrow v'\}$. *BE* denotes the set of basic events, i.e., nodes without any inputs: $BE = \{v \in V \mid \mathsf{Inputs}(v) = \emptyset\}$. *IE* denotes the set of intermediate events together with the top event, i.e., events with at least one input: $IE = \{v \in V \mid \#\mathsf{Inputs}(v) \geq 1\}$. We require that nodes labeled with VOT(k/N) have exactly N inputs: If $Gate(v) = \mathsf{VOT}(k/N)$ then $\#\mathsf{Inputs}(v) = N$.

Example 2 (Status vector and structure function)

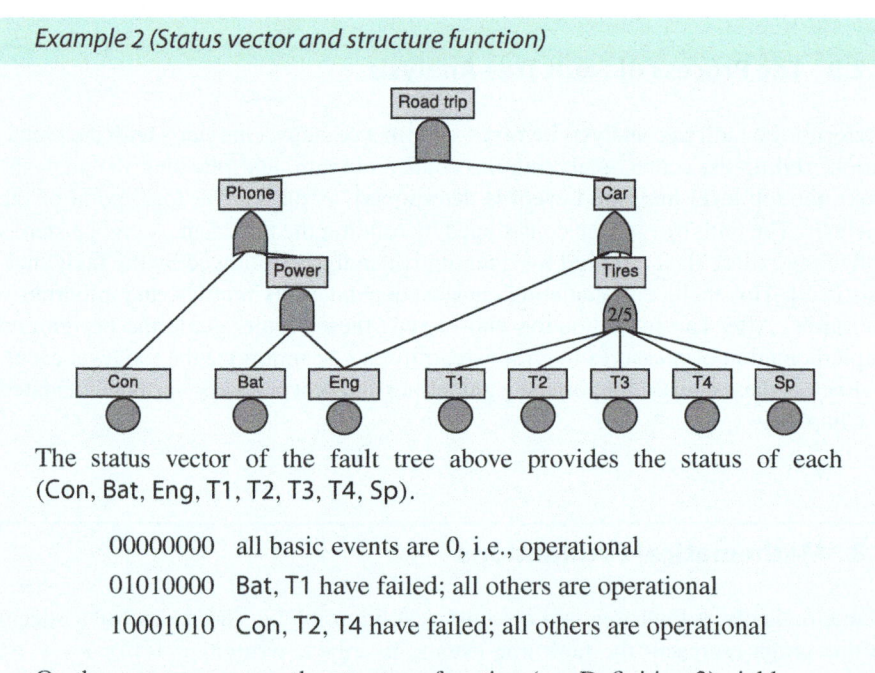

The status vector of the fault tree above provides the status of each (Con, Bat, Eng, T1, T2, T3, T4, Sp).

> 00000000   all basic events are 0, i.e., operational
> 01010000   Bat, T1 have failed; all others are operational
> 10001010   Con, T2, T4 have failed; all others are operational

On these status vectors, the structure function (see Definition 3) yields

$$\Phi_F(00000000) = 0$$
$$\Phi_F(01010000) = 0$$
$$\Phi_F(10001010) = 1$$

2.3.1 Status Vector and Structure Function

A foundational concept is the status vector. This vector sets the status for each basic event e, where 1 means failed and 0 means operational.

Definition 2 A status vector for a fault tree with basic events $e_1, \ldots, e_n$ is a vector of Booleans $\overline{b} = (b_1, \ldots, b_n)$ where

$$\begin{cases} b_i = 0 & \text{denotes that } e_i \text{ is operational} \\ b_i = 1 & \text{denotes that } e_i \text{ has failed} \end{cases}$$

The structure function describes the behavior of the fault tree. For each status vector $\overline{b}$, the structure function Φ_F indicates whether a fault tree F fails on $\overline{b}$, (i.e., $\Phi_F(\overline{b}) = 1$) or whether it remains operational ($\Phi_F(\overline{b}) = 0$). Example 2 illustrates this concept.

Definition 3 The structure function of a fault tree F with basic events $\{e_1, \ldots, e_n\}$ is the function $\Phi_F : \{0, 1\}^n \rightarrow \{0, 1\}$ that takes as input a status vector $\overline{b}$. Again, $\Phi_F(\overline{b}) = 1$ indicates that the fault tree fails; $\Phi_F(\overline{b}) = 0$ means that it is operational.

At times, we use a slight reformulation of Φ_F that takes as input not a status vector, but a *set* of all failed basic events. Thus, the set $\{e_1, e_2, e_5\}$ means that $e_1 = e_2 = e_5 = 1$ and $e_3 = e_4 = 0$, so that $\Phi_F(\{e_1, e_2, e_5\})$ means $\Phi_F(11001)$.

This notation enables defining the notion of cut set as a set of basic events C such that $\Phi_F(C) = 1$. Cut sets are further discussed in Chap. 7.

2.3.2 Obtaining the Structure Function from a Fault Tree Diagram

The structure function Φ_F interprets a fault tree diagram as a Boolean formula. Given a status vector $\overline{b}$, the function $\Phi_F(\overline{b})$ indicates whether F fails on $\overline{b}$. We obtain Φ_F by replacing the gates in the fault tree diagram with their logical symbols, as shown in Example 3. Symbols for Boolean operators are different across disciplines, and summarized in Table 3.1.

Mathematically, the structure function is defined recursively, where an auxiliary parameter E indicates the current event. Thus, $\Phi_F(\overline{b}, E)$ indicates whether event E fails on the status vector $\overline{b}$. Then $\Phi_F(\overline{b})$ is defined as $\Phi_F(\overline{b}, Top)$.

Definition 4 Let F be a fault tree, $\overline{b} = (b_1, b_2, \ldots, b_n)$ a status vector. For a non-basic event E, let $Gate(E)$ be the gate type associated E, being either AND, OR, or VOT(k/N). Further, $e_1, e_2, \ldots, e_m$ are the children of E. Then the structure function

$\Phi_F : \{0, 1\}^n \times V \rightarrow \{0, 1\}$ of F is defined as follows.

$\Phi_F(\overline{b}, E) =$

$$\begin{cases} \Phi_F(\overline{b}, E) = b_i & \text{if } E = e_i \in BE \\ \Phi_F(\overline{b}, E) = \Phi_F(\overline{b}, e_1) \wedge \Phi_F(\overline{b}, e_2) \wedge \cdots \Phi_F(\overline{b}, e_m) & \text{if } Gate(E) = \text{AND} \\ \Phi_F(\overline{b}, E) = \Phi_F(\overline{b}, e_1) \vee \Phi_F(\overline{b}, e_2) \vee \cdots \Phi_F(\overline{b}, e_m) & \text{if } Gate(E) = \text{OR} \\ \Phi_F(\overline{b}, E) = (\sum_{i=1}^{m} \Phi_F(\overline{b}, e_i)) \geq k & \text{if } Gate(E) = \text{VOT}(k/N) \end{cases}$$

Example 3 (Structure function)

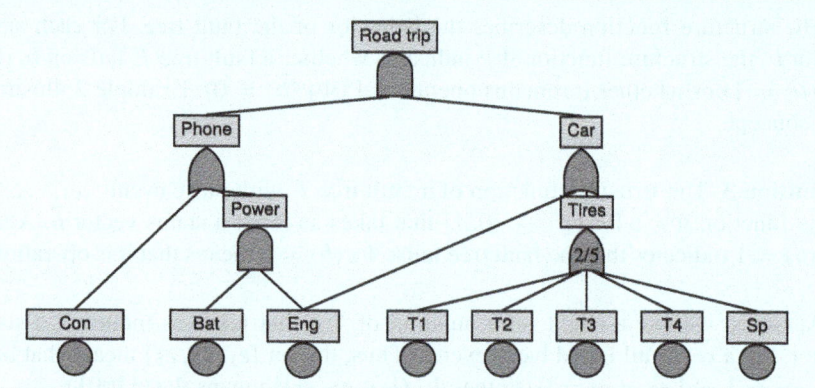

The structure function of the fault tree above is obtained by rewriting the fault tree gates into their logical equivalents, as formulated in Definition 4. Writing $\overline{r} = $ (Con, Bat, Eng, T1, T2, T3, T4, Sp), we obtain

$\Phi_F(\overline{r}) = \Phi_F(\overline{r}, Top))$

$= \Phi_F(\overline{r}, \text{Phone fails}) \wedge \Phi_F(\overline{r}, \text{Car fails}))$

$= (\text{Con} \vee \Phi_F(\overline{r}, \text{Nopower})) \wedge \Phi_F(\overline{r}, \text{Car fails})$

$= (\text{Con} \vee (\text{Bat} \wedge \text{Eng})) \wedge \Phi_F(\overline{r}, \text{Car fails})$

$= (\text{Con} \vee (\text{Bat} \wedge \text{Eng})) \wedge (\text{Eng} \wedge \Phi_F(\overline{r}, \text{Tires}))$

$= (\text{Con} \vee (\text{Bat} \wedge \text{Eng})) \wedge (\text{Eng} \wedge ((\text{T1} \wedge \text{T2}) \vee (\text{T1} \wedge \text{T3}) \vee (\text{T1} \wedge \text{T4}) \vee$

$\qquad (\text{T1} \wedge \text{Sp}) \vee (\text{T2} \wedge \text{T3}) \vee (\text{T2} \wedge \text{T4}) \vee (\text{T2} \wedge \text{Sp}) \vee (\text{T3} \wedge \text{T4}) \vee$

$\qquad (\text{T3} \wedge \text{Sp}) \vee (\text{T4} \wedge \text{Sp}) \vee (\text{T1} \wedge \text{T2} \wedge \text{T3}) \vee (\text{T1} \wedge \text{T2} \wedge \text{T4}) \vee$

$\qquad (\text{T1} \wedge \text{T2} \wedge \text{Sp}) \vee (\text{T1} \wedge \text{T3} \wedge \text{T4}) \vee (\text{T1} \wedge \text{T3} \wedge \text{Sp}) \vee (\text{T1} \wedge \text{T4} \wedge \text{Sp}) \vee$

$\qquad (\text{T2} \wedge \text{T3} \wedge \text{T4}) \vee (\text{T2} \wedge \text{T3} \wedge \text{Sp}) \vee (\text{T2} \wedge \text{T4} \wedge \text{Sp}) \vee (\text{T3} \wedge \text{T4} \wedge \text{Sp}) \vee$

$\qquad (\text{T1} \wedge \text{T2} \wedge \text{T3} \wedge \text{T4}) \vee (\text{T1} \wedge \text{T2} \wedge \text{T3} \wedge \text{Sp}) \vee (\text{T1} \wedge \text{T2} \wedge \text{T4} \wedge \text{Sp}) \vee$

$\qquad (\text{T1} \wedge \text{T3} \wedge \text{T4} \wedge \text{Sp}) \vee (\text{T2} \wedge \text{T3} \wedge \text{T4} \wedge \text{Sp}) \vee (\text{T1} \wedge \text{T2} \wedge \text{T3} \wedge \text{T4} \wedge \text{Sp})))$

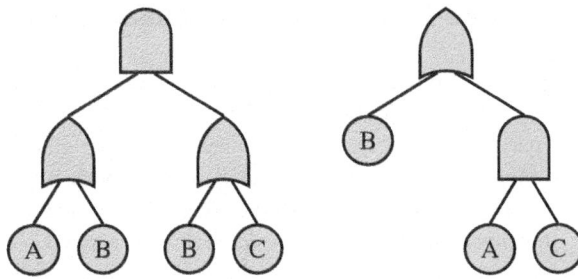

Fig. 2.4 Two equivalent fault trees F and F'

2.3.3 Fault Tree Equivalence

Different fault tree diagrams can be equivalent; that is, they represent exactly the same failure behavior. Since an FT's failure behavior is given by its structure function, two fault trees are equivalent if they have the same structure function. In other words, equivalent fault trees fail on exactly the same set of basic events.

Definition 5 Let F and F' be two fault trees with the same basic events. Then F and F' are equivalent if and only if $\Phi_F = \Phi_{F'}$, i.e., for all $\overline{b} \in \{0, 1\}^n$, we have $\Phi_F(\overline{b}) = \Phi_{F'}(\overline{b})$.

The fault trees in Fig. 2.4 are clearly equivalent: since $\Phi_F(A, B, C, D) = (A \vee B) \wedge (B \vee C)$ and $\Phi_{F'}(A, B, C, D) = B \vee (A \wedge C)$ are the same as Boolean functions: whatever values for A, B, C and D we take, we have $\Phi_F(A, B, C, D) = \Phi_{F'}(A, B, C, D)$.

As expected, this definition yields that the AND-gate with N inputs is equivalent to an VOT(N/N)-gate, and the OR-gate with N inputs is semantically equivalent to a VOT($1/N$) gate. Moreover, a VOTING-gate can be written as an equivalent combination of AND-and OR-gates, for example VOT($2/3$)(E_1, E_2, E_3) = OR(AND(E_1, E_2), AND(E_1, E_3), AND(E_2, E_3)).

Clearly, it can be useful to obtain the smallest fault tree that represents a given structure function. However, this problem is known to be computationally intractable. It has been proven mathematically that this problem is coNP-hard, even limiting to AND- and OR-gates [1]. This means that finding the smallest fault tree is as hard as factorizing an integer into its prime factors, which is for large numbers infeasible with current technology.

Normal forms. As multiple fault trees exhibit the same behavior, normal forms establish a standard format, such that all equivalent trees share the same normal form. The disjunctive normal form is further discussed in Sect. 7.3.

2.3.4 Tree-Shaped Versus DAG-Shaped Fault Trees

Despite their name, fault trees do not have to be proper trees. Proper trees require all branches to be independent. That is, each basic or intermediate event has exactly one parent, i.e., its output is connected to one gate only. We call such fault trees *tree-shaped*. *Shared events,* that is, events with multiple parents as introduced in Sect. 2.1.2, are disallowed in tree-shaped fault trees.

In general, fault trees are *directed acyclic graphs (DAGs)*. When viewing the fault tree as a graph, DAGs allow multiple paths from one event to another, but no loops. Note the lines in fault tree diagrams represent directed edges from a child to its parent, following the direction in which failures propagate. Note that shared events may be depicted either by connecting the shared event to multiple parents, or by repeating the event name multiple times in the fault tree diagram as illustrated in Fig. 2.3.

The distinction between tree- and DAG-shaped fault trees is illustrated in Example 4. The road trip example is not tree-shaped, since Battery fails is a shared event, with both No power and Car fails as parents. The subtrees below Phone fails and Car fails are tree-shaped. The distinction between tree-shaped and DAG-shaped fault trees is important, since tree-shaped fault trees allow for simpler probabilistic analysis methods. However, due to common causes, fault trees are often not tree-shaped.

Example 4 (Tree-shaped versus DAG-shaped fault trees)

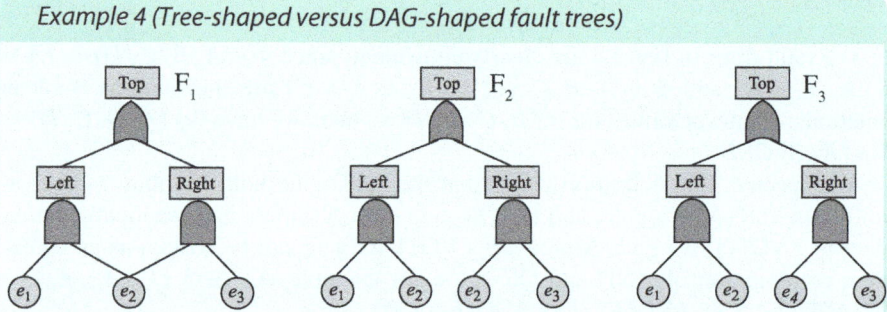

Two DAG-shaped fault trees and a tree-shaped one. In the left-most one, e_2 is a basic event with two parents, hence the fault tree is DAG-shaped. In the middle one, this is the case as well, however, the event name e_2 is replicated in the tree. While more difficult to detect visually, this fault tree is still DAG-shaped, since e_2 has two outputs. The right-most fault tree is tree-shaped.

2.3.5 Fault Trees with Probabilities

Let us assume that each basic event e is equipped with a probability $\mathbb{P}[e]$ that the event e occurs. We can now ask what the probability is that the top event occurs. To do so, we first define the probability of a status vector.

The probability $\mathbb{P}[b_i]$ of a status bit b_i is given by $\mathbb{P}[b_i] = \mathbb{P}[e_i]$ if $b_i = 1$ (i.e., e_i has failed) and $\mathbb{P}[b_i] = 1 - \mathbb{P}[e_i]$ if $b_i = 0$ (i.e., e_i has not failed). This can be written as $\mathbb{P}[b_i] = b_i \mathbb{P}[e_i] + (1-b_i)(1-\mathbb{P}[e_i])$. The probability of a status vector is obtained by multiplying the probabilities of all its status bits, i.e., $\mathbb{P}[\overline{b}] = \mathbb{P}[b_1] \cdots \mathbb{P}[b_n]$, yielding the following definition.

Definition 6 Let $\overline{b} = (b_1, b_2, \ldots, b_n)$ be a status vector. The probability of $\overline{b}$, denoted $\mathbb{P}[\overline{b}]$, is given by

$$\mathbb{P}[\overline{b}] = \prod_{1 \leq i \leq n} b_i \cdot \mathbb{P}[e_i] + (1 - b_i) \cdot (1 - \mathbb{P}[e_i])$$

If we add the probabilities of all status vectors on which the structure function fails (i.e., evaluates to one), then we obtain the probability that the fault tree fails, i.e., the probability that the top event occurs. This is also called the *top event probability*, or shortly the *top probability*.

Definition 7 Let F be a fault tree with n basic events and a structure function Φ_F. The top event probability of F, denoted $\mathbb{P}[F]$, is given by

$$\mathbb{P}[F] = \sum_{\overline{b} \in \{0,1\}^n} \Phi_F(\overline{b}) \cdot \mathbb{P}[\overline{b}]$$

Later in the text, the top event probability of a fault tree F with the top event *Top* is also denoted by p_{Top}. Both notations are equivalent, i.e., $\mathbb{P}[F] = p_{Top}$.

Example 5 (Probabilistic fault tree)

Consider the fault tree **Phone** below, where the basic event probabilities are given below the leaves. The top probability is calculated as shown in the table, by summing the probabilities of all status vectors that result in failure of the top event.

Since this fault tree is tree-shaped, we can also calculate the probability as $p_{Con} + p_{Bat} \cdot p_{Eng} - p_{Con} p_{Bat} \cdot p_{Eng} = 0.2575$.

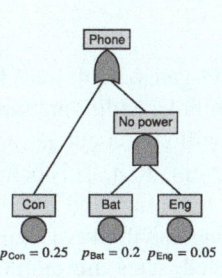

$\overline{b}$	$\Phi_F(\overline{b})$	$\mathbb{P}[\overline{b}]$
000	0	
001	0	
010	0	
011	1	$\overline{p_{Con}} p_{Bat}\, p_{Eng} = 0.75 \cdot 0.2 \cdot 0.05 = 0.0075$
100	1	$p_{Con} \overline{p_{Bat}}\ \overline{p_{Eng}} = 0.25 \cdot 0.8 \cdot 0.95 = 0.19$
101	1	$p_{Con} \overline{p_{Bat}}\, p_{Eng} = 0.25 \cdot 0.8 \cdot 0.05 = 0.01$
110	1	$p_{Con} p_{Bat}\ \overline{p_{Eng}} = 0.25 \cdot 0.2 \cdot 0.95 = 0.0475$
111	1	$p_{Con} p_{Bat}\, p_{Eng} = 0.25 \cdot 0.2 \cdot 0.05 = 0.0025$
		TOTAL $= 0.2575$

Phone — **No power** — Con, Bat, Eng — $p_{Con} = 0.25$ $p_{Bat} = 0.2$ $p_{Eng} = 0.05$

2.4   Additional Fault Tree Symbols

2.4.1   Symbols for Modeling Convenience

The basic fault tree building blocks described above provide enough modeling power to capture any combinatorial structure of failures. In practical applications, the concise and easily understandable structure of fault trees presents a high value on its own. To this end, typical fault trees will include additional symbols, listed in Table 2.2. These do not change the behavior of the fault tree, but are convenient for modeling purposes.

Transfer gates. If a fault tree is too large to fit on one page or screen, then triangles are used to transfer events between multiple FTs to act as one large FT. These triangles are called the *transfer in* and *transfer out* gates.

House events. A *house* event, represents a non-probabilistic event whose state is determined before an analysis. Thus, house events allow portions of a fault tree to be included or not included in analysis. They can aid in analyzing the effects of individual branches of a fault tree. This enables modeling system configurations, operating modes, external factors (weather, season), effects of specific events such as loss of offsite power, or even unavailability of individual components that shall not be treated probabilistic. House events become very practical also for applications built on a fault tree model, such as risk monitoring.

Undeveloped events. An *undeveloped* event, denoted by a diamond, acts similarly to a basic event, as it sits at the bottom of the tree. An undeveloped event indicates that the event could be further refined, and may be used if further resolution is impossible, or it does not have an impact on the resulting analysis. The undeveloped event can also serve as a placeholder for further resolution in the future.

INHIBIT-gate. Finally, the INHIBIT-gate only propagates the input failure in the presence of a condition. That is, for the undesired event to occur, both the input and the condition must be present. For example, a fire only occurs if oxygen is present. Thus, the INHIBIT gate technically works the same as an AND-gate; a different symbol is used for the sake of clarity.

2.4.2   Logic Gates for Negation

The fault trees discussed include AND-, OR-, and VOTING-gates. Fault trees build with these gates are coherent. This means that if the system fails, additional failures will not lead to system success. In other terms, a cut set will stay a cut set, i.e., it will fail the top gate, even if we add arbitrary many basic events to it. This behavior sounds very intuitive for reliable systems. A failure cannot repair a failed system. In spite of this, non-coherent fault trees might simplify modeling and improve legibility of the models. Non-coherent fault trees arise when negated gates are employed, displayed in Fig. 2.5. These require extra caution in analysis and interpretation and

Fig. 2.5 Gates in
non-coherent fault trees

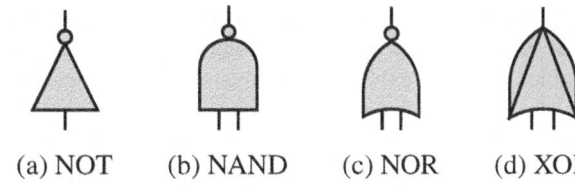

(a) NOT   (b) NAND   (c) NOR   (d) XOR

are discussed in Chap. 15. *In other chapters, we focus on coherent fault trees, unless stated otherwise.*

2.5   Other Methods in Reliability Engineering

Fault tree analysis is not the only method for identifying and quantifying risks. Rather, plethora of methods exist. Similar to [2], we categorize them into three classes.

Text-based methods. Textual approaches provide systematic methods for exploration of components or behaviors in complex systems and list all findings in textual form or a table. Common approaches are Failure Mode and Effects Analysis (FMEA), [3] and Hazard and Operability Study (HAZOP) [4]. Often, an FMEA is conducted in conjunction with the fault tree analysis to identify relevant components and their failure modes which translate into basic events.

Architectural methods. Architectural approaches take as starting point an architectural system model, decomposing a system into a number of interacting components, annotating these with potential risks. Such architectural methods are especially common for systems with large software components, but can be used in any domain with complex system designs. In the nuclear safety domain, the most prominent representative of architectural methods is based on the Figaro modeling language [5]. Section 6.3 details the Figaro method further. Other typical examples are the error annex related to the Architectural Analysis & Design Language AADL [6], AltaRica [7,8], the Safety Analysis Modeling Language (SAML) [9], or Hierarchically Performed Hazard Origin and Propagation Studies (Hip-HOPS) [10]. Another common term for this approach is Model-Based Safety Assessment [11,12].

Domain specific methods. Domain specific approaches are specifically developed or adapted for risk analysis. Apart from fault trees, these include reliability block diagrams [13], event trees [14], Petri nets [15], and Bow tie diagrams [16]. These methods all provide visual means to capture system behavior and offer different analysis possibilities, including Monte Carlo simulations.

Clearly, all methods have their pros and cons. Strengths of fault trees include their systematic construction method, their support for both quantitative and qualitative analysis, the availability of powerful software tools to perform these analyses, and their wide use in practice. The relatively restricted expressive power of fault trees enables efficient exhaustive analysis of very large models. This provides analysts with a possibility to refine a system model to a great level of detail. By this, dependencies and weak points can be discovered with the computational aid that would otherwise rely only on expert judgment. Various organizations, including NASA (National Aeronautics and Space Administration), ESA (European Space Agency), the nuclear industry, and the US federal aviation administration have recognized that a single analytical approach is usually insufficient for effective risk management. Consequently, they suggest a combination of approaches, particularly leveraging top-down and bottom-up analyses.

2.6 A Brief History of Fault Tree Analysis

Fault trees have been developed in the 1960s by H.A. Watson at Bell Laboratories, who worked on it as part of a project with the U.S. Air Force to assess the Minuteman Launch Control System [17], a strategic weapon system for a ballistic missile. Subsequently, Dave Haasl of Boeing recognized their broader utility and applied fault tree analysis to analyze the entire Minuteman Missile System, marking the start of application of fault tree analysis in the aerospace industry, especially within the commercial aircraft division, around 1966 [18].

The wider acknowledgment and application of fault tree analysis (FTA) were boosted by the inaugural System Safety Conference in 1965, sponsored by Boeing and the University of Washington, where several papers on FTA were presented [19]. This conference played a crucial role in sparking global interest in FTA. After its adoption by Boeing, the United States Federal Aviation Administration (FAA) revised airworthiness regulations for transport category aircraft in 1970, introducing failure probability criteria for aircraft systems and driving the widespread adoption of FTA in the civil aviation sector. Subsequently, in 1976, the U.S. Army Materiel Command, a key supplier to the U.S. Army, included FTA in an Engineering Design Handbook on Design for Reliability [20].

Following the military, aviation, and material sectors, the nuclear power industry began utilizing fault tree analysis for designing and analyzing the safety of nuclear power plants, marking a significant phase in the methodology's adoption across industries. By 1975, the U.S. Nuclear Regulatory Commission (NRC) had started using probabilistic risk assessment (PRA) methods, including FTA, with the pioneering study WASH-1400 [21] and significantly expanded PRA research after the 1979 incident at Three Mile Island [22]. This incident led to the release of the *NRC Fault Tree Handbook* in 1981 [23], along with the mandatory use of probabilistic safety assessment under the NRC's regulatory jurisdiction. The nuclear power industry's engagement with FTA notably advanced both the theoretical underpinnings and practical applications of FTA [19].

With widespread acceptance, international standards like IEC 61025 [24] have been established to offer guidance on conducting fault tree analysis, underscoring FTA's integration into global safety and risk management practices [19].

2.7 Concluding Remarks

We like to come back to the quote at the beginning of this book, made by the British statistician George E.P. Box: *All models are wrong, but some are useful.* This holds in particular for fault tree analysis. Although formal evidence is difficult to obtain, it is widely agreed that proper risk assessments, including fault tree analyses, have prevented serious accidents. At the same time, we must realize that fault tree analysis is neither a formal proof nor a guarantee of safety or reliability. This holds in particular for the dependability metrics computed via quantitative analysis: These are not objective numbers, as in Newtonian mechanics, but merely a structured record of the analyst's best understanding. Therefore, these numbers should be handled with care. Decisions based on these numbers reflect informed decisions based on the best information available.

References

1. Bloniarz PA, Hunt HB, Rosenkrantz DJ (1984) Algebraic structures with hard equivalence and minimization problems. J ACM 31(4):879–904. https://doi.org/10.1145/1634.1639
2. Boudali H, Haverkort BR, Kuntz M, Stoelinga M (2007) Best of three worlds: Towards sound architectural dependability models. In: Proceedings of the 8th international workshop on performability modeling of computer and communication systems (PMCCS), pp 45–49
3. Rausand M, Barros A, Hoylan A (2020) Qualitative system reliability analysis. Wiley, Chap 4. Wiley series in probability and statistics. https://doi.org/10.1002/9781119373940.ch4
4. Kletz TA (1999) Hazop & Hazan: identifying and assessing process industry hazards, 4th edn. CRC Press
5. Bouissou M, Bouhadana H, Bannelier M, Villatte N (1991a) Knowledge modeling and reliability processing: presentation of the Figaro language and associated tools. In: Proceedings of the international symposium on computer safety, reliability, and security (SAFECOMP), vol 24, pp 69–75. https://doi.org/10.1016/S1474-6670(17)51368-3
6. Society of automotive engineers (2017) AS5506C: architecture analysis & design language
7. Point G, Rauzy A (2006) AltaRica: constraint automata as a description language. J Européen des Systémes Automatisés 33:1033–1052
8. Arnold A, Griffault A, Point G, Rauzy A (2000) The AltaRica formalism for describing concurrent systems. Fundam Inf 40:109–124
9. Güdemann M, Ortmeier F (2010) A framework for qualitative and quantitative and quantitative model-based safety analysis. In: Proceedings of the IEEE international symposium on high-assurance systems engineering (HASE), pp 132–141. https://doi.org/10.1109/HASE.2010.24

10. Papadopoulos Y, McDermid JA (1999) Hierarchically performed hazard origin and propagation studies. In: Felici M, Kanoun K (eds) Computer safety, reliability and security. Springer, Berlin, pp 139–152. https://doi.org/10.1007/3-540-48249-0_13

11. Lisagor O, Kelly T, Niu R (2011) Model-based safety assessment: Review of the discipline and its challenges. In: Proceedings of the 9th International Conference on Reliability, Maintainability and Safety (ICRMS), IEEE, pp 625–632, https://doi.org/10.1109/ICRMS.2011.5979344

12. Sun M, Gautham S, Ge Q, Elks C, Fleming C (2024) Defining and characterizing model-based safety assessment: a review. Saf Sci 172. https://doi.org/10.1016/j.ssci.2024.106425

13. Modarres M, Kaminskiy MP, Krivtsov V (2016) System reliability analysis, 3rd edn. CRC Press, Chap 4. https://doi.org/10.1201/9781315382425

14. Ericson CA (2005) Event tree analysis. Wiley, Chap 12:223–234. https://doi.org/10.1002/0471739421

15. Leveson NG, Stolzy JL (1987) Safety analysis using petri nets. IEEE Trans Softw Eng SE-13(3):386–397. https://doi.org/10.1109/TSE.1987.233170

16. Center for Chemical Process Safety (2018) Bow ties in risk management. Wiley. https://doi.org/10.1002/9781119490357

17. Watson HA et al (1961) Launch control safety study. Bell Labs

18. Eckberg C (1963) WS-133B fault tree analysis program plan. The Boeing Company, Seattle

19. Ericson CA (1999) Fault tree analysis—a history. In: Proceedings of the 17th international system safety conference, Orlando, Florida, USA, pp 1–9

20. Evans RA (1979) Engineering design handbooks. IEEE Trans Reliab 28(5):421–385

21. US Nuclear Regulatory Commission (1975) Reactor safety study: an assessment of accident risks in US commercial nuclear power plant. WASH-1400. NUREG-75/014

22. Ross D, Murphy J, Cunningham M (1991) NUREG-1150: severe accident risks: an assessment for five us nuclear power plants. Technical report, US Nuclear Regulatory Commission

23. Vesely WE, Goldberg FF, Roberts NH, Haasl DF (1981) Fault tree handbook. Nuclear Regulatory Commission, Office of Nuclear Regulatory Research, U.S

24. International Electrotechnical Commission (2006) IEC 61025: fault tree analysis

Algorithmic Building Blocks

3

This chapter covers several algorithms used in fault tree analysis. It can be skipped on a first read and referred to later as needed. We discuss notation for gates, binary decision diagrams, modularization, and the exponential distribution.

3.1 Notation for Logical Gates and Real Numbers

The notation for logical gates and real numbers varies across different fields. Table 3.1 provides a comparison.

3.2 Binary Decision Diagrams

Binary decision diagrams (BDDs) are a crucial tool in fault tree analysis. They serve both qualitative purposes, such as identifying minimal cut sets, and quantitative purposes, providing effective algorithms to compute dependability metrics. BDDs offer

Table 3.1 Notation for Boolean operators and real numbers

Boolean operator	Mathematics	Engineering
a AND b	$a \wedge b$	a·b, ab
a OR b	$a \vee b$	a+b
NOT a	$\neg a$	$\bar{a}$
Real numbers	$2.5 \cdot 10^{-8}$	2.5E−8

M. Stoelinga et al., *Concise Guide to Fault Tree Analysis*, Computer Science Foundations and Applied Logic, https://doi.org/10.1007/978-3-031-78287-9_3

a compact and efficient way to encode Boolean functions [1,2], which is particularly useful since the structure function of a fault tree is itself a Boolean function that can be represented as a BDD [3,4].

BDDs are widely used across various applications due to their ability to efficiently manipulate logical operators. They are valuable in handling large-scale Boolean formulas in fields such as hardware verification, software analysis [5], satisfiability checking, and model checking. Furthermore, highly optimized software packages [6] are available to support the necessary BDD computations, making them a powerful tool in both theory and practice.

This section provides several concepts that are used throughout the book. This chapter can be skipped on the initial read and revisited when needed.

Definition 8 A BDD is an acyclic graph that encodes a Boolean function f : $\{0, 1\}^n \rightarrow \{0, 1\}$. The leaves of this graph are labeled with the truth values 0 and 1, and often depicted as squares. All other nodes v are labeled with x_i, i.e., one of the input variables of f. Such nodes have two successors:

- the left child v_L represents the function in case $x_i = 0$ and is usually depicted with a dashed line;
- the right child v_R represents the function $x_i = 1$, usually depicted with a solid line.

Since we utilize BDDs to encode a fault tree's structure function $\Phi : BE \rightarrow \{0, 1\}$, in our case, each BDD node v gets labeled with a basic event $x_i \in BE$.

> *Example 6*
>
> Figure 3.1a shows a fragment of the road trip fault tree. Its structure function is given by $\Phi(\mathsf{Con, Bat, Eng}) = \mathsf{Con} \vee (\mathsf{Bat} \wedge \mathsf{Eng})$. Each circle in Fig. 3.1b represents a BE, and has two children: a 0-child that determines the system status if this BE has not failed, and a 1-child for if it has. The leaves of the BDD indicate whether the entire fault tree has failed.

The values of the function f can be found back in its BDD representation by starting at the top and following the corresponding path. In Fig. 3.1, the value $\Phi(010)$ is found by taking respectively the left/dashed, right/solid, left/dashed edges in Fig. 3.1 This path leads to the leaf labeled 0. Indeed, $\Phi(010) = 0 \vee (1 \wedge 0) = 0$.

Ordered and reduced BDDs. To exploit the full power of BDDs, fault tree analysis methods work with BDDs that are ordered and reduced: *Reduced* BDDs do not allow duplicate subtrees, leading to compact BDDs, and efficient analysis. Clearly, when two subtrees are equivalent we need only one of these. Two nodes v and v' are equivalent if they represent the same Boolean function. Concretely, not only the top node, but each node in a BDD represents a Boolean function $f_v : \{0, 1\}^m \rightarrow \{0, 1\}$, where $m < n$, over a subset of the variables $x_1, \ldots, x_n$ that appears in the

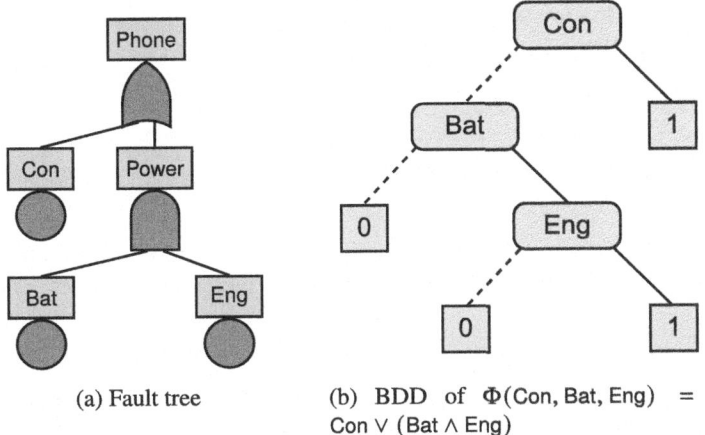

(a) Fault tree

(b) BDD of Φ(Con, Bat, Eng) = Con $\vee$ (Bat $\wedge$ Eng)

Fig. 3.1 Fault tree and its BDD representation. Dashed lines indicate a 0 and solid lines a 1 for the corresponding BE. The values of Φ are given by following the corresponding path. For example, $\Phi(010)$ is found by taking respectively the left/dashed, right/solid, left/dashed edges. This path leads to the leaf labeled 0. Indeed, $\Phi(010) = 0 \vee (1 \wedge 0) = 0$

subtree below v. Now, nodes v and v' are equivalent if they represent the same function, i.e., $f_v = f_{v'}$.

The BDD in Fig. 3.3c is reduced[1], while the one in Fig. 3.3a is not. One can reduce a non-reduced BDD by removing duplicate subtrees. In this way, we obtain an equivalent BDD that is smaller.

3.2.1 Deriving a BDD from a Fault Tree

A key technique to construct a BDD from a Boolean formula f is the so-called *Shannon expansion* of f [1,7]. Starting from the variable order $x_1 < \ldots < x_n$, one constructs the top node v_1 of the BDD by expanding

$$f(x_1, x_2, \ldots, x_n) = (\neg x_1 \wedge f(0, x_2, \ldots, x_n)) \vee (x_1 \wedge f(1, x_2, \ldots, x_n))$$

The Shannon expansion is named the if-then-else method: if $x_1 = 0$, then the function equals $f(0, x_2, \ldots, x_n)$ else equals $f(1, x_2, \ldots, x_n)$.

From the Shannon expansion, the BDD is created in the following way: First, BDD-node v_1 is created, which is labeled by x_1. Then the left child of v_1 is connected to the BDD obtained from the Shannon expansion of $f(0, x_2, \ldots, x_n)$. Similarly, the right child is connected to the BDD obtained from the Shannon expansion of

---

[1] Officially, reduced BDDs have only one node labeled 0 and one node labeled 1. We leave multiple copies for clarity in the pictures.

$f(1, x_2, \ldots, x_n)$. Recursively applying this expansion until all variables have been converted into BDD nodes yields a complete BDD.

Example 7

An example of the process of Shannon expansion to create a BDD is shown in Fig. 3.2.

- We begin in step (a) with the structure function of a fault tree, here $\Phi =$ Con $\vee$ (Bat $\wedge$ Eng). We arbitrarily choose variable ordering Con $<$ Bat $<$ Eng.
- Now, we apply the Shannon expansion for the variable Con, namely:

$$\Phi(\text{Con}, \text{Bat}, \text{Eng}) = (\neg\text{Con} \wedge \Phi(0, \text{Bat}, \text{Eng})) \vee (\text{Con} \wedge \Phi(1, \text{Bat}, \text{Eng}))$$
$$(\neg\text{Con} \wedge (0 \vee (\text{Bat} \wedge \text{Eng}))) \vee (\text{Con} \wedge (1 \vee (\text{Bat} \wedge \text{Eng})))$$

This is shown in step (b), where the top node shows the split on variable Con, with the dashed/left branch going to the $\neg$Con side, and the solid/right branch going to the Con side. This process is continued in step (c) for variable Bat and in step (d) for variable Eng.
- Now that all variables of the structure function have been converted into nodes of the BDD, the structure function can be evaluated to obtain the values in the leaves of the BDD, as shown in step (e).
- Finally, we reduce the BDD by collapsing duplicate subtrees. In Fig. 3.3, we see that the leftmost decision node (on Eng) has value 0 in both children, so this subtree can be collapsed into just a 0-value leaf, as shown in (f). Similarly, the entire subtree on the right can be reduced to a leaf with a value of one.

Figure 3.4 depicts the BDD for the entire road trip example.

3.2.2 BDDs in Fault Tree Analysis

BDDs have two important roles in fault tree analysis.

Computation of the top event probability. The top event probability can be computed by decorating the BDD with the failure probabilities of all the basic events. For each BDD-node v labeled with a basic event e, we decorate the right child with probability p_e and its left child with probability $(1 - p_e)$. Then the top probability is computed as the sum of the probabilities of all paths in the BDD leading to the BDD leaf labeled 1. Here, the path probabilities are obtained by multiplying the probabilities of all edges in the path. This procedure is detailed in Sect. 10.4.

Computation of minimal cut sets. The BDD representation of a fault tree gives a compact encoding of all cut sets in the fault tree, where each path leading to a 1-leaf

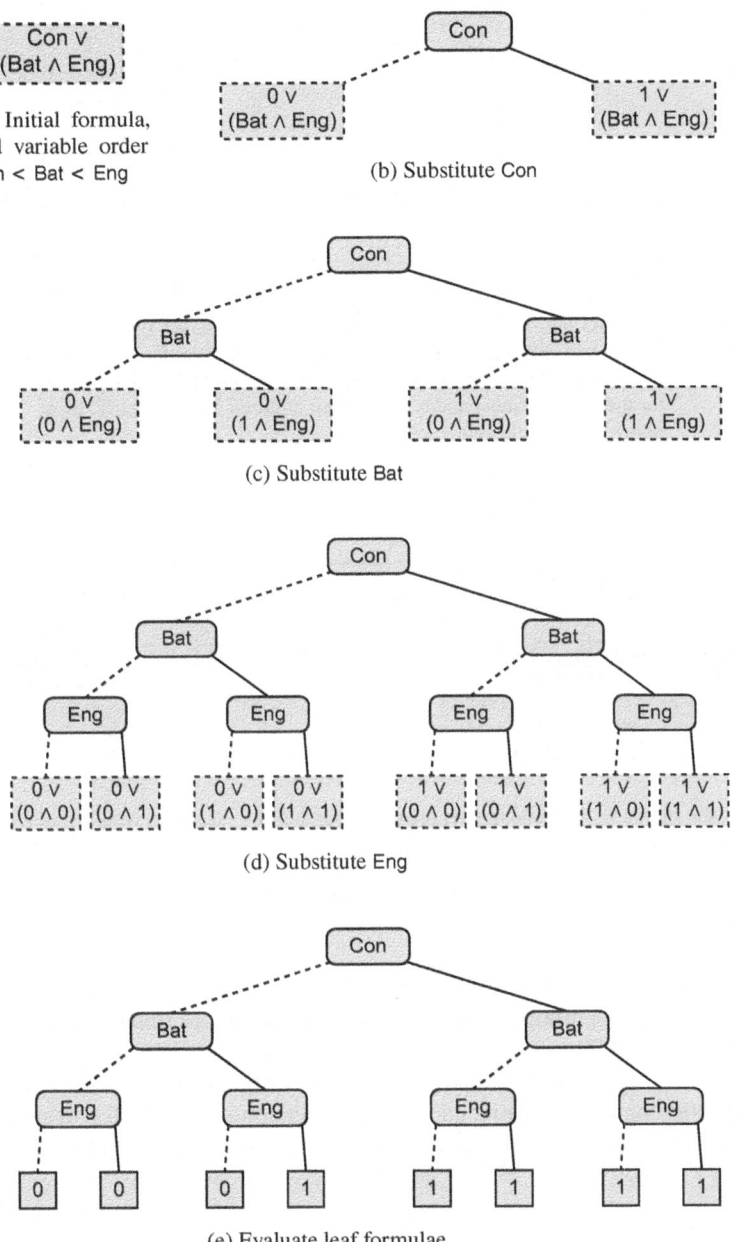

Fig. 3.2 Stepwise execution of Shannon expansion to create a BDD with variable order Con < Bat < Eng. Start is the structure function Φ = Con ∨ (Bat ∧ Eng) and the lowest variable is Con. Each node expands into two children: The left/dashed child sets the variable to 0 and the right/solid child to 1

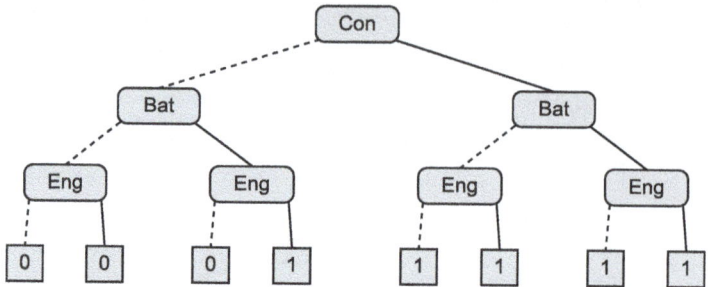

(a) BDD before reduction: the leftmost node labeled Eng has two symmetric children, each leading to the 0-leaf. Therefore, the value of Eng does not matter, so this node can be removed. The same holds for the two rightmost nodes. Since their children are symmetric, these nodes can be removed.

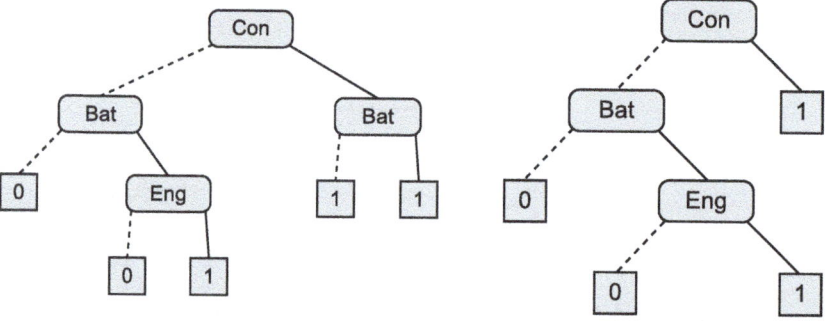

(b) Now the rightmost Bat node has symmetric children and can be removed

(c) Final results: reduced BDD

Fig. 3.3 Reducing the BDD by identifying identical subtrees

represents a cut set. However, these cut sets need not be minimal. To obtain a BDD representation of all minimal cut sets, a modified structure function must be used to generate the BDD. To compute the top event probability, either version of the BDD (obtained via the original or modified structure function) can be utilized. For the probabilistic computation, one can use both versions of the BDD, using either the original or the modified structure function. Details are presented in Sect. 2.3.4.

3.3 Modularization

Modularization is a crucial technique for efficient fault tree analysis.

Recall that fault trees can either be tree-shaped or DAG-shaped, see Sect. 2.3.4. Many algorithms are fast on tree-shaped fault trees, and far less efficient on DAGs. This holds, for example, for methods computing the minimal cut set list and for

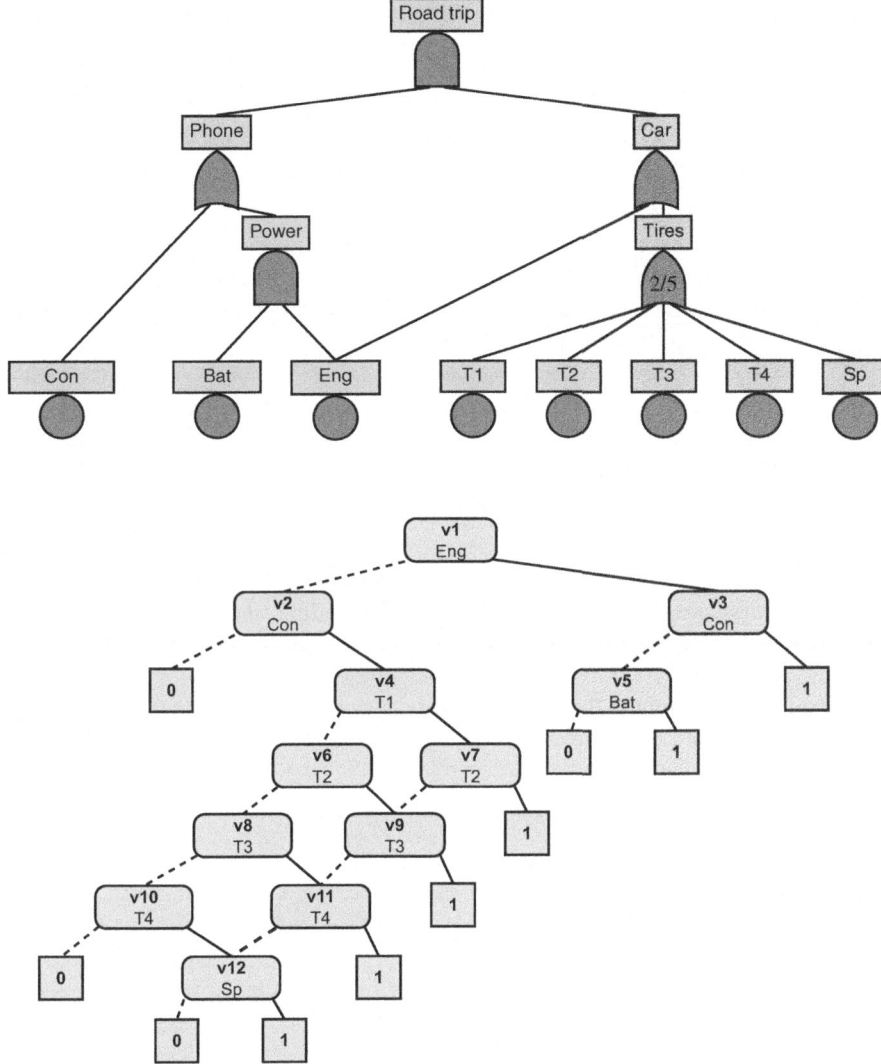

Fig. 3.4 BDD for the road trip fault tree

computing the top probability. These can be computed via a linear-time bottom up procedure on tree-shaped fault trees, whereas DAG-shaped fault trees require methods, such as BDDs, that are exponential in the worst case.

Therefore, splitting up a fault tree in tree-shaped and DAG-shaped parts can significantly improve efficiency, allowing faster algorithms to be deployed on the tree-shaped parts of the fault tree, while less efficient algorithms for DAGs are only deployed where needed. This technique is called *modularization* [8].

A module represents an independent subtree of the fault tree, which can be analyzed separately, independently of the rest of the fault tree. That is, nodes in this subtree have no connections leading to nodes outside the subtree.

Definition 9 For a node $E \in V$, let F_E denote the subtree under E. A node E is a *module* if every path from any node of F_E to *Top* passes through E.

The literature often excludes basic events from being considered as modules. We do include basic events as modules, as this simplifies the formulation of several properties. In particular, a fault tree is tree-shaped if and only if all nodes are modules.

Example 8

The road trip fault tree has modules: all basic events, the top event, and the intermediate event Tires. None of the other intermediate events (specifically, Phone, Power, and Car) are modules. In each case, the basic event Eng has a path to the top that bypasses the intermediate event in question.

As emphasized throughout the book, modules often support compositional (a.k.a. modular) computation. That is, if a node E is a module, then its subtree can be analyzed separately, and the result of the subtree F_E can be further propagated up the tree.

For example, to compute the top probability, we can first compute the probability p_E of F_E, and ignore the entire structure of F_E, as if we created a new basic event $\tilde{v}$ whose probability equals p_E.

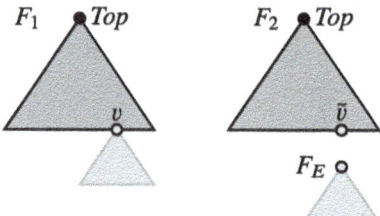

3.3.1 Fault Tree Pre-processing

A fault tree created by an analyst will typically have a structure that supports easy understanding of the model by other analysts or reviewers. This structure may be less suited for analysis. Several rewriting steps transform a fault tree into a structure that is faster to analyze.

Typical processing steps are simplifying the fault tree's structure function through the laws of Boolean algebra [9]. For example, simplifying $(e_1 \vee e_2) \wedge (e_1 \vee e_3)$ to $e_1 \vee (e_2 \wedge e_3)$ does not only simplify the formula, it also transforms a DAG-structured tree into a tree-shaped tree.

Another step is to increase the modular structure for a fault tree. For example, assume that an intermediate event $E = \text{AND}(E_1, E_2, \ldots, E_n)$ has as modules $E_2, \ldots, E_n$, but E_1 is not a module. Then E can be rewritten as $E = \text{AND}(E_1, E')$, where $E' = \text{AND}(E_2, \ldots, E_n)$ is a module.

3.4   The Exponential Probability Distribution

One of the most common failure time distributions is the (negative) exponential distribution. This distribution carries one parameter λ, called the *failure rate*, describing the average number of failures per time unit. The unique property of the exponential probability distribution is that this failure rate is time-independent: After three years, the component fails as fast as after one hour. The exponential distribution can be realistic during the normal life of a component; see the bathtub curve in Fig. 11.3. Moreover, exponential distributions are fundamental for understanding other distributions.

Exponential failure times. We consider a basic event e whose failure time is governed by an exponential distribution with parameter λ. To define what this means, we consider the point in time at which the first failure of e occurs. We denote this time point by T_e.

$$T_e = \min\{t \in [0, \infty) \mid e \text{ is in a failed state at time } t\}$$

Since failures occur at random, T_e is a random variable. Its *failure (time) distribution* is given by F_e, i.e., $F_e(t)$ is the probability that e fails before time t.

$$F_e(t) = \mathbb{P}[\text{Event } e \text{ fails before time } t] = \mathbb{P}[T_e < t]$$

Technically, F_e is the cumulative distribution function of T_e. Saying that e is exponentially distributed with parameter λ means that

$$F_e(t) = 1 - e^{-\lambda t}$$

Exponential as a consequence of time independence. We can derive the cumulative distribution function from the time-independence property as follows. Let α be the probability that e does not fail within a single unit of time, i.e., $\alpha = \mathbb{P}[T_e > 1]$.

Since the failure speed is constant, the probability for e to not fail in two time units is

$$\mathbb{P}[T_e > 2] = \mathbb{P}[T_e > 1] \cdot \mathbb{P}[T_e > 1] = \alpha^2$$

Indeed, for e to not fail in two time units, it must not fail in the first time unit, and neither in the second time unit. With the speed of failure being time independent, both probabilities are equal to α. Similarly, the probability for e to not fail in the first half time unit is $\mathbb{P}[T_e > \frac{1}{2}] = \alpha^{\frac{1}{2}} = \sqrt{\alpha}$. In general, we have $\mathbb{P}[T_e > t] = \alpha^t$. This is also called the *survivor* time, since the component has to survive the first t time units. Substituting $\lambda = -\ln \alpha$, i.e., $\alpha = e^{-\lambda}$ yields

$$\mathbb{P}[T_e > t] = \alpha^t = (e^{-\lambda})^t = e^{-\lambda t}$$
$$F_e(t) = \mathbb{P}[T_e < t] = 1 - \mathbb{P}[T_e > t] = 1 - e^{-\lambda t}$$

Its probability density function can easily be computed through differentiation.

$$f_e(t) = F'_e(t) = \lambda e^{-\lambda t}$$

As stated, λ represents the average number of failures per time unit. Then the average time before a failure occurs, i.e., the mean time to failure, is given by $\frac{1}{\lambda}$: Three failures per time unit means an average failure time of $\frac{1}{3}$.

$$\mathbb{E}[T_e] = \int_0^\infty t f_e(t) dt = \int_0^\infty t \lambda e^{-\lambda t} dt = \frac{1}{\lambda}$$

Example 9

Assume a basic event e with a component failure time that is exponentially distributed with parameter λ, written $T_e \sim \exp(\lambda)$. That is, $F_e(t) = 1 - e^{-\lambda t}$. Then probability that e is operational at time t is given by

$$\mathbb{P}[X_e = 1] = \mathbb{P}[e \text{ has failed within time } t] = \mathbb{P}[T_e < t] = 1 - e^{-\lambda t}$$
$$\mathbb{P}[X_e = 0] = \mathbb{P}[e \text{ has not failed within time } t] = \mathbb{P}[T_e > t] = e^{-\lambda t}$$

For example, if a component has a failure rate of 6 failures per year, then we take $\lambda = 6$. In this way we obtain

- $\mathbb{E}[T_e] = \frac{1}{\lambda} = \frac{1}{6}$, i.e., the mean time to failure is 1/6 years, i.e., 2 months.
- $\mathbb{P}[T_e < 0.5] = 1 - e^{-3} \approx 0.95$, i.e., the probability to fail within the first half year is 95%.

OR-gate with exponential distribution. A noteworthy property is that the failure time of an event $E = \mathrm{OR}(e_1, e_2)$ of two exponentially distributed events is itself exponentially distributed with parameters $\lambda_1 + \lambda_2$.

This can be seen as follows. We consider the probability for E to fail after time t. This is exactly the case when both e_1 and e_2 fail after time t.

$$\mathbb{P}[T_E > t] = \mathbb{P}[T_{e_1} > t \wedge T_{e_2} > t]$$
$$= e^{-\lambda_1 t} e^{-\lambda_2 t}$$
$$= e^{-(\lambda_1 + \lambda_2) t}$$

Clearly, $\mathbb{P}[T_E > t] = 1 - e^{-(\lambda_1+\lambda_2)t}$, i.e., T_E is exponentially distributed with parameter $\lambda_1 + \lambda_2$.

If an event is an AND of two exponentially distributed events, we do not get an exponential distribution for the combination, as shown here:

$$\mathbb{P}[T_E < t] = \mathbb{P}[T_{e_1} < t \wedge T_{e_2} < t]$$
$$= (1 - e^{-\lambda_1 t})(1 - e^{-\lambda_2 t})$$
$$= 1 - e^{-\lambda_1 t} - e^{-\lambda_2 t} + e^{-(\lambda_1+\lambda_2)t}$$

Exponential versus geometric distribution. If, for an exponential distribution with parameter λ, we only study the failure behavior at discrete time points $t = 0, 1, 2, \ldots$, then we obtain a *geometric distribution* for the failure time T'_e of e. At each discrete time point $t = 0, 1, 2, \ldots$, the component remains operational with probability α, and fails with probability $(1 - \alpha)$. As before, $\alpha = e^{-\lambda}$. Then the probability to fail at exactly time t is given by

$$\mathbb{P}[T'_e = t] = \alpha^{t-1}(1 - \alpha)$$

Thus, we still have $\mathbb{P}[T'_e > t] = \alpha^t$ and $\mathbb{P}[T'_e \leq t] = 1 - \alpha^t$. The geometric distribution has mean $\frac{1}{1-\alpha}$, while the exponential distribution has mean $\frac{1}{\lambda} = \frac{1}{-\ln \alpha}$, i.e.,

$$\mathbb{E}[T'_e] = \frac{1}{1 - \alpha}$$
$$\mathbb{E}[T_e] = \frac{1}{\lambda} = \frac{1}{-\ln \alpha}$$

The reason for this discrepancy is that, in the exponential distribution, e can fail in between time points $0, 1, 2, \ldots$, but in the geometric distribution it cannot. This makes the component fail slower than the exponential distribution: Consider $t = 1$. For example, the geometric distribution cannot fail before time 1, whereas the exponential distribution can:

$$\mathbb{P}[T'_e < 1] = \mathbb{P}[e \text{ fails before time 1, geometric distribution}] = 0$$
$$\mathbb{P}[T_e < 1] = \mathbb{P}[e \text{ fails before time 1, exponential distribution}] = 1 - \alpha.$$

As a consequence, the average failure time is also shorter for the exponential distribution than for the corresponding geometric distribution. A detailed comparison between the geometric and the exponential distributions is given in [10].

References

1. Akers SB (1978) Binary decision diagrams. IEEE Trans Comput C-27(6):509–516. https://doi.org/10.1109/TC.1978.1675141
2. Bryant RE (1986) Graph-based algorithms for Boolean function manipulation. IEEE Trans Comput C-35(8):677–691. https://doi.org/10.1109/TC.1986.1676819
3. Schneeweiss W (1985) Fault tree analysis using binary decision diagrams. IEEE Trans Reliab 34(5):453–457
4. Rauzy AB (1993) New algorithms for fault tree analysis. Reliab Eng Syst Saf 40(3):203–211. https://doi.org/10.1016/0951-8320(93)90060-C
5. Whaley J, Avots D, Carbin M, Lam MS (2005) Using datalog with binary decision diagrams for program analysis. In: Yi K (ed) Programming languages and systems. Springer, Lecture Notes in Computer Science, vol 3780, pp 97–118. https://doi.org/10.1007/11575467_8
6. van Dijk T, van de Pol J (2017) Sylvan: multi-core framework for decision diagrams. Int J Softw Tools Technol Transf 19(6):675–696. https://doi.org/10.1007/s10009-016-0433-2
7. Shannon CE (1938) A symbolic analysis of relay and switching circuits. Electr Eng 57(12):713–723
8. Dutuit Y, Rauzy AB (1996) A linear-time algorithm to find modules of fault trees. IEEE Trans Reliab 45:422–425. https://doi.org/10.1109/24.537011
9. Platz O, Olsen J (1976) Faunet: a program package for evaluation of fault trees and networks. Technical Report, Report No. 348, Research establishment risk, Roskilde, Denmark
10. Hermanns H (2001) Construction and verification of performance and reliability models. Bull EATCS 74:135–154

Part II

Fault Tree Analysis as a Process

The Role of Fault Tree Analysis in Decision Making

4

4.1 Purpose of Fault Tree Analysis

The primary objective of fault tree analysis is to facilitate accountable decision-making regarding the risks associated with the system being studied. These risks can encompass the system design, operations, and dismantling, and cover various dependability aspects, such as safety, reliability, and availability. Subsequent actions may involve a range of intervention measures to decrease the failure probability by design changes, such as increasing redundancy, or by improving the reliability of critical components or by improving operation and maintenance. Associated dependability measures might help prioritizing investments or improve asset performance by measures related to design, training or operation.

Understanding and documentation. Even without a quantitative analysis, the fault tree provides a systematic overview of system vulnerabilities and potential failure scenarios, capturing complex interactions between components in the presence of failures. The mere activity of creating fault trees by systematically breaking down high-level risks into their contributing factors is often advantageous, revealing vulnerabilities and areas needing attention. These insights help early-stage and proactive risk mitigation strategies. Moreover, fault tree diagrams serve as documentation, ensuring traceability and preserving these insights for future use.

Compliance and certification. Governments and certification bodies often require verification that systems meet dependability requirements, especially for safety-critical systems like nuclear plants, airplanes, and self-driving cars. These requirements, often expressed as dependability metrics, ensure reliability and safety standards, see Sect. 2.2. For example, Dutch flood protection barriers must have a failure frequency below $1 \cdot 10^{-4}$ per year. Regulatory requirements for major nuclear accidents range between $1 \cdot 10^{-4}$ and $1 \cdot 10^{-6}$ per year, with a tendency to push this limit even lower to $5 \cdot 10^{-7}$ per year. Data centers may guarantee 99.99% uptime.

M. Stoelinga et al., *Concise Guide to Fault Tree Analysis*, Computer Science Foundations and Applied Logic, https://doi.org/10.1007/978-3-031-78287-9_4

Fault tree analysis techniques efficiently compute conservative estimates for these metrics, making compliance a key purpose.

Design and operational decisions. The ultimate goal of fault tree analysis is to improve system dependability, by devising measures that prevent the top event from happening or reduce its probability. With numerous potential measures available, selecting (cost-)effective ones is challenging. Fault tree analysis can support this process.

Such measures can affect the design and operations. Typical design measures include: using more reliable components, using redundancy and diversification, fault isolation, separation or shielding, spare parts management, training of personnel, condition monitoring of system components, and a complete redesign. Clearly, to profit from design decisions, the fault tree analysis must be performed at design time.

Typical operational measures include strategies for testing and inspection, monitoring, maintenance, and replacement. In a power plant, a question can be if the conditions are still safe enough to operate. If a failure occurs while some part of the system is under maintenance, safety may be at stake. A re-analysis of the fault tree under these specific conditions facilitates risk-informed decisions. For the road trip example, such operational strategies can be to check the tire profiles each year, inflate the tires after each trip and replace the tires every six years.

Other areas related to operations that benefit from fault tree analysis include design of operator training, accident management and control of operational practices. High probability accident scenarios might get additional focus when planning training of plant operators. Procedures for accident management will include actions motivated by results of fault tree analysis. It might help to rank actions according to their importance in a given accident situation. Operational practices could be made after the design. However, since operations have a large effect on the system dependability, it is advisable to consider operational practices already during the design phase.

These design and operational decisions are typically supported by quantitative analysis using cut set metrics, via dependability metrics and/or importance measures described above. Subsequently, one can compute the effect of measure taken, for example by implementing them in the fault tree model and re-calculating relevant dependability metrics. Chapters 10 and 13 elaborate on how to use these metrics in design decisions.

Diagnosis and monitoring. If a system failure has occurred, fault trees can help to understand how this failure happened [1]. If an existing tree is available, then one can trace back the failure's root causes. The idea is to start at the top event and walk down the tree: in case of an AND-gate, one checks if indeed all subcauses have occurred; in case of an OR-gate, one has to investigate which of the children has occurred. If problems occurred that do not appear in the tree, one needs to revise the tree—and decisions based on this tree. If no fault tree is available, one identifying the root cause can be constructed, by developing only those events that have occurred.

Monitoring refers to the continuous evaluation of risk. For example, if an incident happened that did not lead to a system failure, re-analysis of the fault tree can assess the significance of the incident. Importance measures are often deployed here. An

Table 4.1 Quantitative techniques supporting a given purpose of FTA

	Cut sets	Cut set probabilities	Dependability metrics	Importance measures
Understanding	x	x		
Compliance		x	x	
Design and operational decisions	x	x	x	x
Diagnosis	x	x	x	x

important question is if the diagnostic analysis and monitoring should give rise to additional measures. On the one hand, prevention of failures and safety hazards is important; on the other hand, over-managing risks leads to expensive and impractical solutions.

Table 4.1 outlines common methods that facilitate a specific objective of fault tree analysis. These should be taken as guidelines, not as absolute rules.

4.2 Fault Tree Analysis in Standardization, Probabilistic Safety Analysis, and Reliability Engineering

Probabilistic safety assessment and reliability engineering. Fault tree analysis is an integral part of the wider field of RAMS: Reliability, Availability, Maintenance, Safety. In probabilistic safety assessment, the focus on fault trees is used to reduce safety risks. In a safety case, the top level event is a safety-related hazard, such as exposure-to-death, injury, loss, or damage. The objective of reliability engineering is similar, except that the top level event is often related to non-catastrophic failures.

ISO Standards recommending fault tree analysis. ISO standards play a crucial role in ensuring quality, safety, and efficiency across industries, by establishing internationally recognized norms for their assessment. In safety or business-critical markets, clients often demand their customers to be ISO-certified. Various ISO standards require, recommend or use fault trees.

- *IEC 61508* is a basic functional safety standard applicable to all industries. The *IEC 61508* standard addresses the functional safety of electrical, electronic, and programmable electronic systems. Fault tree analysis is recommended for evaluating safety-related systems.
- The *IEC 61511* standard sets out practices in the engineering of systems that ensure the safety of an industrial process through the use of instrumentation.

- *ISO 26262* standardizes analysis of functional safety for automotive systems. It requires quantified metrics to be calculated for safety-related systems. Fault tree analysis is recommended, especially for common mode failures.
- The *ISO 14971* standard pertains to the application of risk management to medical devices.
- The *ISO 31000* standard provides guidelines for risk management across various sectors. While it does not specifically mention fault tree analysis, the methodology aligns with the principles and goals of fault tree analysis.
- Finally, the *ISO/IEC 27001* standard outlines the requirements for information security management systems, where fault trees can be deployed for security risk analysis.

Some other standards include:

- IAEA standard [2] for nuclear probabilistic safety assessment models states requirements on fault trees for modeling system failures in accident sequences.
- The NUREG-0492 handbook [3] codifies and systematizes the fault tree approach to system analysis.
- The ASME/ANS RA-S-1.1-2024 standard [4] is co-developed by the American Society of Mechanical Engineers (ASME) and the American Nuclear Society (ANS). The 2024 edition focuses on Level 1/Large Early Release Frequency Probabilistic Risk Assessment (PRA) for Nuclear Power Plant operations, aimed at assessing the core damage frequency (CDF).

Other bodies recommending fault tree analysis are the Federal Aviation Administration [5], International Atomic Energy Agency [2] and the U.S. Nuclear Regulatory Commission [6]. Fault tree analysis itself is standardized by the International Electrotechnical Commission (IEC) as standard IEC 61025 [7].

References

1. Hurdle EE, Bartlett LM, Andrews JD (2008) System fault diagnostics using fault tree analysis. Proc Inst Mechan Eng Part O: J Risk Reliab 221(1):43–55. https://doi.org/10.1243/1748006XJRR6
2. International Atomic Energy Agency (2024) Development and application of level 1 probabilistic safety assessment for nuclear power plants. No. SSG-3 (Rev. 1) in Specific Safety Guides, IAEA, Vienna
3. Vesely WE, Goldberg FF, Roberts NH, Haasl DF (1981) Fault tree handbook. Nuclear Regulatory Commission, Office of Nuclear Regulatory Research, U.S
4. American Society of Mechanical Engineers (ASME), American Nuclear Society (2024) RA-S-1.1-2024 Standard. https://www.asme.org/codes-standards/find-codes-standards/standard-for-level-1-large-early-release-frequency-probabilistic-risk-assessment-for-nuclear-power-plant-applications

5. Federal Aviation Administration (2000) System safety handbook
6. US Nuclear Regulatory Commission, Office of Nuclear Regulatory Research (1981) Fault tree handbook. Technical report, U.S. Nuclear Regulatory Commission
7. International Electrotechnical Commission (2006) IEC 61025: Fault tree analysis

The Process of Fault Tree Analysis

5

5.1 Steps in a Fault Tree Analysis

The process of fault tree analysis follows a structured procedure, making the decisions more consistent, thorough, and reliable. This process closely follows the ISO 31000 standard on risk management [1]. The steps are outlined below and depicted in Fig. 5.1. While presented as a sequential process, these steps are usually carried out in an iterative process, involving several iterative cycles, going back to previous steps if needed. Documentation of all steps is of crucial importance as it ensures traceability, learning, communication, compliance, and continuous improvement. Chapter 6 details several steps on a concrete case from the nuclear industry.

5.1.1 Preparation

1. Set the objective of the analysis. Establishing the objective of the fault tree analysis is a crucial starting point for the analysis process. This involves identifying the questions to be answered and dictates the expertise, level of detail, time frame, and data required for the analysis. The purposes in Chap. 4 (understanding, certification, design of improvement measures, diagnostics) can help to set the objective. If quantitative analysis is to be performed, then the dependability metrics of interest must be specified, as well as the role of these metrics: do they serve as requirements, target value, or as information.

2. Set scope and resources. As usual, a good preparation is key to the success of fault tree analysis. We recommend at least the following steps:

a. Set the scope of the analysis, system boundaries, and aspects taken into account.
b. Gather all the information available: design documents, failure records, documentation from manufacturers and clients, etc.

© The Author(s), under exclusive license to Springer Nature Switzerland AG 2026
M. Stoelinga et al., *Concise Guide to Fault Tree Analysis*, Computer Science Foundations and Applied Logic, https://doi.org/10.1007/978-3-031-78287-9_5

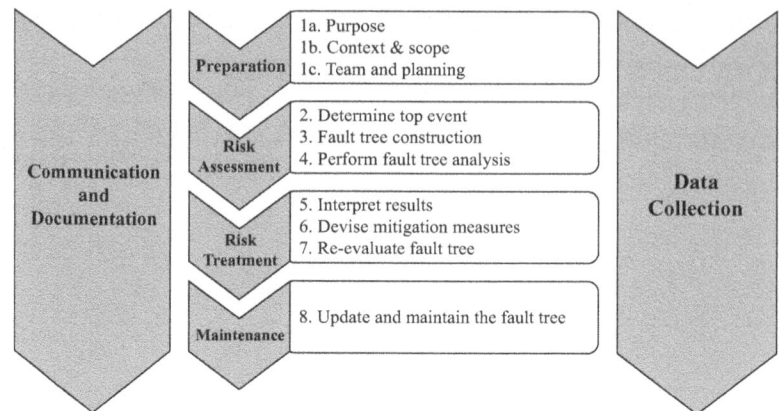

Fig. 5.1 Steps in the fault tree analysis process

(a) *System description*: describing context of the system, its inner workings, basic operation, and components.
(b) *Failure definitions*: an exact description from a system authority (safety manager, mission manager) of when the system is considered to have failed.
(c) *Operational profile*: describing the (intended) use of the system, such as usage intensity, frequency, accepted maintenance intervals, availability of maintenance teams before and during operation.
(d) *Assumptions about the system*: collect any assumptions about the system.

c. Form a team and allocate roles. Typical roles include: An *FTA contractor*, who sets the goals and scope, secures the budget and other resources, A *coordinator*, who oversees and coordinates the whole process; *Domain experts,* who know the product and who are versed in its failure modes; *Contextual specialists*, such as customers, suppliers, asset managers and maintenance engineers also bring valuable information to the table; *FTA reviewers* review periodically the quality of the fault tree analysis.
d. Discuss planning. The required time frame strongly depends on the purpose. For example, design evaluations are most useful when conducted along with the design, enabling timely improvements.

3a. Determine the top level undesired events. The risk assessment phase starts by identifying all undesirable events that require analysis, called *pivotal events*. Each pivotal event becomes the top event of a fault tree, which may overlap with other trees. Other sources, such as hazard analyses, FMECAs, or known mishaps, can help to find pivotal events.

Different operating modes or usage scenarios often lead to multiple top events. For example, in the case of a fire protection system, people typically want to know if (1) the system will not produce unexpected events under normal circumstances, i.e., when there is no fire, and (2) if the system will function in the event of a fire. As a fire system operates differently in standby and active modes, these are analyzed by separate but potentially overlapping fault trees.

A clear definition for each pivotal event is essential, as it forms the basis of the fault tree diagram. Finding a clear definition for the top event should not be underestimated and can be a point of debate. Even in our small road trip example, one could argue that the top event Road trip is somewhat limited. If there is an extremely reliable phone, such as a satellite phone with a robust battery, the occurrence of the top event will be rare. However, it raises the question of whether it truly constitutes a highly reliable road trip if the car frequently breaks down but a phone is always on hand. Therefore, one may study the Car fails event as a (secondary) top event, investigating if the reliability of the car is acceptable. In this case, no separate tree is required, but the results are computed for two events.

3b. Determine analysis goals and the metrics to be computed. For qualitative analysis, the analysis goal can be to check if single points of failure exist and to compute the list of all cut sets. If a quantitative analysis will be conducted, it is crucial to specify the metrics that will be calculated, reflecting the performance indicators of interest. These metrics determine the data required for the analysis. Chapter 11 provides an overview of the most common metrics related to the various purposes of the analysis.

5.1.2   Risk Assessment

4a. Construct the fault tree. Once the top event has been defined, the actual tree construction can start. The initial step in building a fault tree involves conducting a so-called Failure Modes and Effects Analysis (FMEA, also called Failure Modes, Effect and Criticality Analysis, FMECA). This analysis screens the system design for components that are relevant for system failures. Each component can fail or be unavailable when needed in many different ways. A specific way in which component fails to perform its intended function is called a *failure mode*. Failure modes are influenced by many factors, not only the construction of the component, but also its function within the system, and its technical or physical conditions. For example, a centrifugal pump may fail to start, fail during operation, or be inaccessible due to scheduled preventive maintenance. The pump can also fail to function due to a faulty or missing incoming signal, a failing power supply, or insufficient cooling. Depending on the scope, it could be determined that failures caused by a flood, earthquake, or fire are not relevant to this evaluation. FMEAs identify these failure modes and their effects on the system functionality. Whereas fault tree analysis is a top-down method, starting from the system level failure and refining until its root causes, FMEA is a bottom-up method: it starts from the root causes and examines its effect at the system level.

In some cases, fault tree diagrams can be automatically obtained, if a high-level system description is available in a formalized form that contains all relevant information from a reliability point of view. Chapter 6 summarizes several approaches in this direction. In this case, the effort of creating the fault tree is shifted to the creation of the high-level description.

4b. Validate the fault tree. Ensuring that the fault tree faithfully reflects reality is important, as mistakes can significantly skew the analysis results. An incremental validation process, which involves checking smaller subtrees throughout the construction, helps detect errors early and simplifies debugging. Consequently, validation is not merely a post-construction step, but rather an ongoing procedure that runs in parallel to the fault tree's development. The following steps are useful.

1. Determine minimal cut sets of the fault tree, verifying their reasonableness against the top event. Use the design documentations or FMEA.
2. Identify paths set in the fault tree, validating their accuracy akin to Step 1.
3. Identify relevant low-order faults (i.e., intermediate events), obtain their cut and path sets, and validate them as in Steps 1 and 2.
4. Assess the probabilities of cut sets and their contributions for reasonableness.
5. Evaluate the probabilities of intermediate faults for reasonableness, comparing them with past data.
6. Assess the probability of the top event, compare it with similar assessments, and consider the probability of extreme events realistically.

In addition, reviewers are expected to review the resulting fault tree.

5. Equip all basic events with failure data. If quantitative analysis will be performed, then failure data is required for all basic events in the fault tree. Such failure data should be expressed as failure probabilities or failure rates, depending on the type of analysis required.

Data collection is very time consuming. As indicated in Fig. 5.1 and Sect. 5.1.6, the data collection process is not performed at a single point in time, but rather as a continuous process during operations.

6. Perform fault tree analysis. Next, the actual fault analysis should be performed, calculating the goals and metrics established in Step 3b for the fault tree constructed in Step 5.

5.1.3   Risk Treatment

7. Interpret the results. After performing the analysis, interpret the results and communicate the relevant information to the necessary stakeholders. The results of any analysis depend on the quality of the input data and the assumptions made during the analysis. As such, it is important to view the results as a starting point for further investigation and validation, rather than a definitive conclusion.

8. Devise mitigation measures. Based on the findings of the fault tree analysis, appropriate mitigation measures and improvements are designed and implemented. These measures may either eliminate or decrease the likelihood of an undesired event, or reduce its impact. Numerous improvement measures exist: *Design measures* change the system design. These include the use of more reliable components (i.e., with lower failure rates), redundancy (i.e., having several components with the same functionality, so that one can take over if one fails), diversification (where the redundant components are of different types or form different manufacturers, making them less prone to common cause failures such as production or design errors), spare management (i.e., keeping spare parts on shelves to quickly replace broken components), fault isolation, separation or shielding (all blocking failures to propagate through the system), and monitoring the condition of system components and training if human factors are involved. The most drastic measure is a complete redesign. *Operational measures* are performed while the system is running. Typical measures include strategies for testing and inspection, maintenance, and replacement.

5.1.4 Maintenance of the Fault Tree

8. Monitor progress and maintain the fault tree. Lastly, it is crucial to continuously monitor the performance of these improvements and regularly update the fault tree to reflect any changes in system design, operating conditions, or component performance. This involves not only revising the fault tree structure but also updating the failure data to ensure the tree remains accurate and valuable to the organization.

5.1.5 Communication and Documentation

Clear records enable stakeholders to understand decisions, learn from mistakes, and comply with legal requirements. The traceability of results, along with the documented steps and considerations that led to them, offers deeper insight into potential risks and their causes, and serves as a foundation for future analyses.

5.1.6 Data Collection

Collecting reliable data is a crucial task that can consume a significant portion of the analysis effort and should therefore not be underestimated. Many equipment suppliers provide failure data for their components, often tailored to different operational modes.

First, it is essential to determine whether the system is being analyzed at a specific point in time or if its failure evolution is being examined over an extended period of time. In the former case, we need a failure probability for each basic event. In the latter case, where we perform a time-dependent analysis, we work with time-dependent failure distributions, which can for instance be given by failure rates. Further, for components that are repairable, we need repair times.

An important decision is to establish a baseline for the system's health condition. In particular, when analyzing an existing system, it is crucial to consider to what extent the current state influences your data. Again, explicit documentation is required for traceability.

Time-independent failure analysis. The time-independent probabilistic model studies the failure behavior of a system at a single time point. This probability can refer to a specific point in time or an unspecified moment in the future. In such cases, it is necessary to determine the failure probabilities for each basic event, indicating the likelihood of failure occurring at that time point.

Time-dependent failure analysis. The time-dependent model examines how a system's failure behavior evolves over time, requiring failure distributions that describe the time-based failure behavior of each basic event. Typically, failure rates are used to represent the average time to failure for a component, serving as the parameter for an exponential failure distribution.

When components are repairable, it becomes relevant to study system availability, which refers to the percentage of time the system is operational. There are two scenarios to consider: observable failures, which are immediately noticeable when they occur, and unobservable or hidden failures, which are not immediately apparent. Inspections are necessary to detect hidden failures and repair them before they cause a system failure. For example, in a car, a flat tire is an observable failure, while a malfunctioning airbag is typically unobservable but can be identified during routine inspections.

In both cases, repair time distributions are essential, as they indicate how long it takes to restore the system to operational status. Additionally, for hidden failures, inspection times or intervals are necessary to determine how frequently inspections should be conducted. Moreover, the repairability of components can vary: some components can be repaired when failures are revealed, others may have unrevealed failures, and some may not be repairable at all.

Common metrics for repairable systems include mean time to failure (MTTF), mean time to repair (MTTR), and mean time between failures (MTBF), where $MTBF = MTTF + MTTR$. Calculating these metrics requires the same data used for determining availability, and notably, MTTF is also applicable in unrepairable cases.

Importance measures. Importance measures quantify the impact of a basic event on a probabilistic dependability metric. Several importance measures exist, working with either failure probabilities or failure distributions.

5.2 Guidelines for Building Fault Trees

5.2.1 Assumptions

This book considers static fault trees. Being built from Boolean logic, these trees are conceptually very simple. This brings advantages to model updates, reviewing and quality control, and the interpretation of results, as well as to the analysis. Current tools can handle huge fault trees with tens of thousands of gates and leaves. A major part of the conceptual simplicity stems from the static nature of fault trees: To understand whether the system has failed and the top event has been reached, it only matters which basic events have failed. The failure order does not matter.

Application of fault trees requires several assumptions:

- A system consists of components with clearly defined failure modes.
- To conduct quantitative analysis, it is essential to describe the failure behavior of each basic event using probabilistic models. In addition, we must be able to quantify the parameters for each of these models, obtaining the failure probability, failure frequency, or failure intensity of each component and its failure modes.
- Logical structure of failure propagation through the system can be established and expressed by Boolean logic. Moreover, negations play a specific role that differentiates fault trees from Boolean formulas.
- Top event(s) can be identified on a reasonably high level as system failures or even some surrogate measures for loss of life, property or damage to environment.

5.2.2   Guidelines for the Construction of Gates

Several rules and guidelines exist for constructing high-quality fault trees [2, 3].

- *Assume no miracles.* When a malfunctioning component happens to prevent or repair a system failure, it is seen as an exceptional event and should not be incorporated into the fault tree.
- *Complete-the-gate.* It is good practice to define all inputs to a gate before any further development of any input is undertaken.
- *No gate-to-gate.* Intermediate events should be explicitly defined and gates should not be directly connected to other gates.
- *No wires/pipes.* Exclude wiring and piping faults unless significantly impactful.
- *No out-of-design.* Omit out-of-design conditions unless justified.
- *OR.* Exclude inputs to OR-gates with significantly lower probabilities than others.
- *Address CCFs.* Address contributors to common cause failure (CCF) for identical active components.
- *Handle complexity.* Use component fault trees (c.f. Sect. 16.1.4) to break down a large fault tree into manageable pieces.

5.2.3   Guidelines for Basic Events

A fault tree's basic events can be of various nature. They can be internal and external events, like component failures, human errors, and environmental conditions. Consulting with subject matter experts, and/or review of historical data, incident reports, and maintenance records, is essential at this stage of the analysis.

1. *Human failures*. A relevant human factor in the road trip example is the ability to mount the spare wheel if one of the primary tires fails. In fact, one can argue that the spare tire fails not only if the tire fails, but also if the car travelers do not manage to correctly mount the spare wheel on the car, or there is no equipment available for the replacement. We can include this situation in the fault tree by further refining the basic event Spare fails by an OR-gate with children Spare tire fails physically, Mount failure and No equipment.
2. *Software failures*. Even though failure data is difficult to obtain, software failures do have a place in fault trees. Most importantly, software is getting an ever increasing role in all our systems.
3. *Cyberattacks*. Basic events often model unintended failures. However, malicious attacks become increasingly important. Such attacks can also be modeled as basic events in a fault tree. However, defects due to cyberattacks have a formalism of their own, namely attack trees. Attack trees and fault trees are similar, but subtly different. In particular the interpretation of the OR-gate in attack trees is often different from fault trees. Combinations of attack trees and fault trees also exist, in several variants. Section 16.1.2 discusses attack trees.
4. *Common cause failures*. Common cause failures are important as they often have a dominant effect on the system performance. It is good practice to include these as basic events in the fault tree. Chapter 14 discusses the modeling of common cause failure in detail.

When to stop refining the tree. An important decision in the fault tree construction is when to stop refining. Usually, one refines until the level where decisions are made. If quantitative analysis will be performed, then one refines until the level where data is available.

On demand versus in-operation failures. Two kinds of failures exist: On demand and in-operation failures. A component like a motor or a generator can fail to provide the required function if it either fails to start, or if it fails during the time when it is expected to operate. The first failure represents an on-demand failure and can be naturally modeled by a single probability – counting the portion of occasions when we requested the component to start and it did not. The time of the failure is *now*, at the moment of request. The second failure represents an in-operation failure. The physical meaning of this event assumes that the component started operating and after some time in operation, but before the required duration of its function was over, the component stopped to work correctly. This could evoke evolution of the system in time, but the mathematical meaning of this event reduces to a mere probability with which the component did not manage to operate throughout the complete mission time. Ultimately, both failures become probabilities.

Basic event naming schemes. Many organizations have strict conventions for naming the basic events. For example, basic events should include the area code, system locator code, component function code, function identifier, sequence code, component type code, and failure mode code.

5.3 Software Tools Supporting Fault Tree Analysis

Several commercial and academic tools exist to support the fault tree drawing and analysis process. The list below presents a short and non-exhaustive overview.

RAM Commander. ALD Software Ltd. (Advanced Logistics Development) produces an FTA program as part of its RAM Commander toolkit [4]. It features standard minimal cut set based analysis as well as Binary Decision Diagrams. Apart from typical dependability measures, RAM Commander can calculate importance and sensitivity information. This program can also automatically generate FTs from FMECAs, FMEAs, or RBDs. It supports product trees, event trees, Monte Carlo simulations of RBD and other types of dependability analyses beyond fault trees.

CAFTA. CAFTA (Computer Aided Fault Tree Analysis) [5] is a tool developed by EPRI (Electric Power Research Institute) for Probabilistic Safety Assessment. It supports single- and continuous-time FTs, including non-coherent fault trees, CCFs, minimal cut set based analysis with post-processing features, importance, sensitivity and uncertainty analyses. As a tool for nuclear safety analyses, it supports building, traceability, reviewing and analyzing of large event and fault tree models. It interfaces other modules for specific types of analyses, such as seismic hazards, fire and flooding, and human reliability analysis.

IsoGraph's Reliability Workbench. The Reliability Workbench, developed by Isograph, includes a fault tree tool FaultTree+ [6]. It integrates fault tree modeling and analysis with event trees, FMEA and Markov analysis. The minimal cut set based analysis engine offers common cause failures, importance and sensitivity analysis and uncertainty simulations. It can handle also non-coherent fault trees with success quantification as an option. Reliability Workbench is widely used in the automotive industry.

ReliaSoft BlockSim. BlockSim module [7] of the ReliaSoft suite allows modeling system reliability, availability and maintainability by Reliability Block Diagrams as well as fault trees and Markov diagrams. Analysis either uses minimal cut sets for static models or Monte Carlo simulations for dynamic ones. The fault tree modeling includes also gate and event types from dynamic fault trees.

Relyence Fault Tree. Relyence tool suite developed by Relyence Corporation includes a module for fault tree modeling and analysis [8]. Users can automatically generate fault trees from FMEA or enter them manually. The analytical engine calculates standard reliability and availability measures, importance and minimal cut sets.

RiskSpectrum PSA. RiskSpectrum [9] is a complete platform for Probabilistic Safety Assessment, developed by RiskSpectrum AB and used by a vast portion of the commercial nuclear power plants worldwide. It facilitates creation, updates and analysis of complex event and fault tree models. Apart from minimal cut set generation and quantification, importance, sensitivity, uncertainty and time-dependent analysis, it offers functionality such as tools for model completion, minimal cut set tracing, seismic hazard analysis, and human reliability analysis. The analysis engine can quantify event tree success, repairs of components, and estimate the minimal cut set list probability by Binary Decision Diagrams.

SAFEST. The SAFEST tool set [10] is designed for static and dynamic fault trees, offering an easy-to-use graphical user interface and great flexibility in the fault tree parameters.

SAPHIRE. Idaho National Laboratory develops Systems Analysis Programs for Hands-on Integrated Reliability Evaluations (SAPHIRE) [11] that facilitates building large event and fault tree models with common cause failure modeling, minimal cut set generation, importance and sensitivity analyses. It interfaces other programs for quantification or human reliability analysis.

References

1. International Organization for Standardization (2018) ISO 31000: Risk management – guidelines. ISO Standard. https://www.iso.org/standard/65694.html
2. Stamatelatos M, Vesely W, Dugan JB, Fragola J, Minarick J, Railsback J (2002) Fault tree handbook with aerospace applications. Office of safety and mission assurance NASA headquarters
3. Andrews J, Lunt S (2012) An introduction to fault tree analysis. Tutorial
4. Advanced Logistics Development (Accessed Jul 2024) Fault Tree Analysis (FTA) Software. http://aldservice.com/en/reliability-products/fta.html
5. EPRI (Accessed Jul 2024) CAFTA. http://www.epri.com/abstracts/Pages/ProductAbstract. aspx?ProductId=000000000001015514
6. Isograph (Accessed Jul 2024) FaultTree+. www.isograph.com/software/reliability-workbench/fault-tree-analysis/
7. ReliaSoft (Accessed Jul 2024) BlockSim. www.reliasoft.com/BlockSim/index.html
8. Relyence (Accessed Aug 2024) Fault Tree. https://relyence.com/wp-content/uploads/2024/03/Relyence-Fault-Tree-Brochure.pdf
9. RiskSpectrum AB (Accessed Jul 2024) RiskSpectrum software. https://www.riskspectrum. com/
10. DGB Technologies (2023) SAFEST—your source for safety solutions. https://www.safest. dgbtek.com/. Accessed on April 2024
11. Smith CL, Wood ST (2021) Systems analysis programs for hands-on integrated reliability evaluations (SAPHIRE) version 8, vol 1–7, no 7039 in NUREG/CR Series, U.S. Nuclear Regulatory Commission. https://www.nrc.gov/reading-rm/doc-collections/nuregs/contract/cr7039/index.html

Creating Fault Tree Models

6.1 Fault Tree Modeling Process

The following section describes the fault tree modeling process in an industrial setup. Fault trees represent a formalism or a tool for reliability, availability, maintainability and safety (RAMS) assessment of a system, whose type can range from an industrial plant such as a nuclear power plant or a chemical plant, over airplanes, submarines and cars, to specific missions involving a technical device such as a flight to Mars. Analysis of each of these systems will want to answer different questions motivated by different concerns.

Nuclear power plants represent a highly regulated industry with statutory requirements on demonstrating safety. There is a clear public interest in assuring the safety during the complete lifecycle, translated to specific regulations, regulatory bodies and processes. State authorities require a satisfactory answer to these concerns to issue a license for operation. Building a probabilistic safety assessment model which includes fault trees has become a standard part of this assessment. This model does not have to be built and maintained solely for the purpose of licensing. In many cases, this probabilistic model is also used for applications other than licensing [1].

An important area is the evaluation of improvements or design changes. The exhaustive nature of the fault tree analysis provides the possibility to systematically identify and rank safety weaknesses. This can help designers to focus on modifications that are most relevant from a safety perspective. Also, several alternative designs can be compared against various safety objectives.

A probabilistic safety assessment model of a nuclear power plant can be after some adaptations used also for real-time monitoring of the plant risk level during operation. The state of the plant is determined by many factors such as the configuration of systems, planned equipment outage due to preventive maintenance, unplanned equipment outage due to a failure, plant operating state or even the weather conditions. As these factors change over time, the risk of a major accident leading to a reactor core damage or a risk of shutting down the plant changes, too. Knowing this

© The Author(s), under exclusive license to Springer Nature Switzerland AG 2026

M. Stoelinga et al., *Concise Guide to Fault Tree Analysis*, Computer Science Foundations and Applied Logic, https://doi.org/10.1007/978-3-031-78287-9_6

risk may influence operating decisions to satisfy regulatory requirements, technical specifications for operation, but also to maximize the production of electricity.

Other 'risk informed' applications of a nuclear power plant model include control of operating practices, training, maintenance and testing, accident management strategies, or evaluation of safety relevant events that occur during plant life time. As two examples, training might focus on accident scenarios calculated from the probabilistic model that have a high contribution to a specific risk. The process where inspectors evaluate their findings, called Significance Determination Process, might quantify effects of an event on the plant risk level by re-quantifying the fault tree model.

For a spaceflight mission, a critical concern is avoiding a loss of mission or its degradation preventing it from achieving its goals. The failure probability in all phases of the mission should be minimized. A fault tree based model can reveal weaknesses in the design and it can guide improvements so that the mission success probability is increased in an efficient way.

A probabilistic analysis of a type of railway rolling stock, such as a passenger car, aims at estimating the portion of the vehicle life time when it is unavailable for commercial operation. The passenger car manufacturer or later its owner want to maximize its utilization to ensure high returns on their investments. The analysis might lead to identifying design aspects that can cause unacceptable downtime.

Similar economic concerns motivate fault tree analysis of utilities or technical equipment for production such as gas turbines. The model captures scenarios that could lead to a degraded production capacity. It allows the manufacturer to estimate the portion of time when the production capacity drops below a certain threshold. By this, we can estimate the availability of the analyzed device and its expected production over a given period of time.

In general, one could say that a fault tree based model facilitates risk-informed decisions. It ranges over granting a license for operation, choosing one out of the alternative design proposals, optimizing the maintenance program of technical assets, or deciding on the next operation action for a running plant. Each of these purposes shapes the fault trees by different formulation of failures, different criteria for screening out or abstracting parts of the design, and different demands on model features. It also influences when in the engineering process it is meaningful to complement the design by a RAMS model based on fault trees.

6.2 Obtaining Fault Trees: Nuclear Safety Example

The probabilistic safety assessment process of a nuclear power plant, nowadays codified in both national and international norms and regulatory requirements [1], starts with defining safety goals. The ultimate aim of avoiding harm to people and environment obtains a more specific, technical form by defining surrogate goals. We want to obtain a structured and verifiable argument that the reactor core gets damaged only in exceptional cases. Typical safety goals range between the frequency of once in ten thousand operating years and once in one hundred thousand years. There is a tendency to decrease this frequency, sometimes going down to five occurrences in one million years.

In the next step, a probabilistic safety assessment project defines the scope. Shall we analyze only internal events, only a certain class of internal events, shall we assess risks connected to flooding or an earthquake? This gives us a set of so-called initiating events. These events mark the start of a potential accident scenario. They bring the plant out of its normal operation and might require stopping the plant: stopping the turbine and the chain reaction inside of the reactor.

Each power plant contains a number of safety systems that shall mitigate the effect of the initiating event. These systems might succeed or fail, resulting in different evolutions of the accident, so-called accident sequences. Some of these sequences avoid the unwanted consequence, e.g., core damage, and reach a stable and safe condition of the plant. Other sequences might lead to a core damage. We want to study the latter group of sequences. For this, we have to understand failure logic of the relevant systems within the power plant. This failure logic will be captured by fault trees.

6.2.1 System Design Description

Design of a system is typically defined by a schematic system engineering description using a graphical representation developed for a certain type of systems, such as Piping and Instrumentation Diagrams (P&ID) for hydraulic systems or a Single-Line Diagram for electrical systems. Figure 6.1 shows a schematic of a very simple hydraulic system that we will use to illustrate the fault tree modeling process. This system is intended to transport fluid from the source to the sink. It consists of four redundant trains. Each train contains three components: a motor-driven pump, a motor-operated valve and a check valve. Components are denoted by a schematic picture with a name. Names of motor-driven pumps end with the suffix 'MP', names of motor-operated valves end with the suffix 'MV' and names of the check valves end with 'CV'. Links between pictures denote piping between components. The arrow on each link shows the direction of the fluid flow.

6.2.2 Identifying and Defining Basic Events

Each component is designed to behave in certain ways. It can happen that it fails to behave as intended—a component failure occurs. For example, a pump could stall and stop transporting the fluid, or a check valve could be blocked and not allow fluid through, or could fail to close and allow fluid to flow in the reverse direction.

There are several steps that one must go through before drawing fault trees. First, we need to know which basic failures, e.g., failures of pieces of equipment, components, support systems or operator failures, are relevant for the analysis. There are

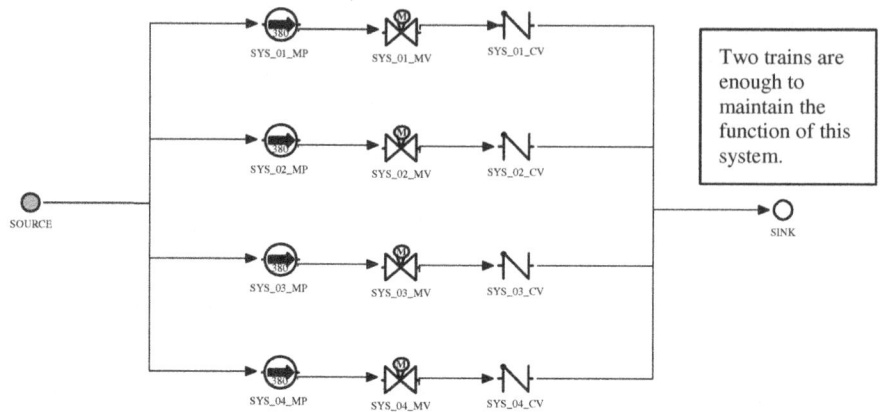

Two trains are enough to maintain the function of this system.

Fig. 6.1 A schema of a sample cooling system with four redundant trains

different methods for identifying basic failures that will translate to basic events in fault trees.

Especially for larger and more complex systems, the first step in building a fault tree is so-called Failure Modes and Effects Analysis (FMEA). This analysis screens the system design for components relevant for system failures. Each of these components can fail or simply be unavailable when needed because of different reasons, such as maintenance. We identify relevant ways, so-called failure modes, in which it can fail to provide the expected functionality. These depend on the component itself, its role in the system and other technical or physical circumstances. As an example, a centrifugal pump can fail to start, it can fail in operation, it can be unavailable because of an ongoing preventive maintenance, it can fail to function because of a faulty or missing incoming signal, no power supply or insufficient cooling. Depending on the scope, we might conclude that a failure because of a flood, earthquake, or fire are not relevant for this assessment. A motor operated valve might fail to open or fail to close, depending on which function do we expect from it. It might be in an incorrect position at the beginning of the accident and this position might not be detected. Also, the signaling or power supply might fail.

This process might uncover dependencies between systems. For instance, the power supply and the digital instrumentation and control turned out to be relevant for the failure modes of the centrifugal pump and the motor operated valve. If we have not included the power supply system in the assessment, we have to add it now and proceed with its Failure Modes and Effects Analysis. The type of relevant dependencies is also determined by the scope of the assessment. For a fire or flooding analysis, there might be dependencies between components given by the fact that they are physically located in the same room.

A typical safety assessment process will include other types of analyses prior or parallel to drawing fault trees. Success of certain accident mitigating barriers might depend on operator actions, which are subject of a Human Reliability Analysis

(HRA). Here we list relevant operator actions, study their dependencies to various factors such as the accident progression, crew state and skills, stress level, possible previous operator actions, and, based on this, quantify probabilities of human failures. A seismic analysis represents another example, quantifying building or component failures as a response to a seismic activity of a certain strength. Deterministic analyses identify and quantify events relevant for accident sequences leading to radioactivity releases such as physical phenomena in the containment.

Finally, we have to populate basic events with failure data. Nuclear power industry invests a considerable effort into collecting and evaluating failures from the operating experience. This statistical process provides parametric characterization of failure mode probabilities. Sometimes, other methods are used such as engineering judgment, manufacturer data obtained from tests, mathematical models for failure estimation such as various methods in Human Reliability Analysis, or deterministic calculations for physical phenomena.

6.2.3   Drawing Fault Trees

Assume that we have a design scheme of a system, a system engineering description such as a Piping and Instrumentation Diagram (P&ID) or a Single-Line Diagram. We have screened out parts irrelevant for the safety/reliability/availability study and identified failure modes of the remaining components. There is a small number of places in this scheme where we could 'measure' (sub-)system failure by a simple test. We can specify that a (sub-)system failure could be determined by 'measuring' a certain process or quality. If this place corresponds to the output of a system, it might correspond to a top event. Otherwise, it becomes an intermediate event.

For example, there is no flow of cooling water at the end of a redundant train in a cooling system. Or there is no voltage at a certain place in an electric power supply system. Consider the system described in Fig. 6.1. The node SINK denotes the output of this system. This node can be translated to a top gate. Success criteria for this system determine the type of the gate.

The process of developing a fault tree traverses the system design scheme from the identified failure point—top event—backward and always answers the question: "What has to happen in the immediate upstream component (or a block of components) so that the unwanted failure at this place results from it." We ask for a Boolean combination of potential immediate causes of the analyzed failure. Failure modes of components correspond to basic events. A summary of failures further away upstream is denoted by a gate whose inputs develop those upstream failures. This process ends when we have processed the whole system design scheme, more precisely, its safety relevant projection and we have reached basic events.

Assume that two trains are sufficient to maintain the system function. Then the system top gate will be a 3-out-of-4 gate with the gates representing each of the redundant trains as inputs. For several places, the logic under the system top gate connects them by expressing which combinations of failures will lead to system unavailability. If we had a separate place for each train in our example system and

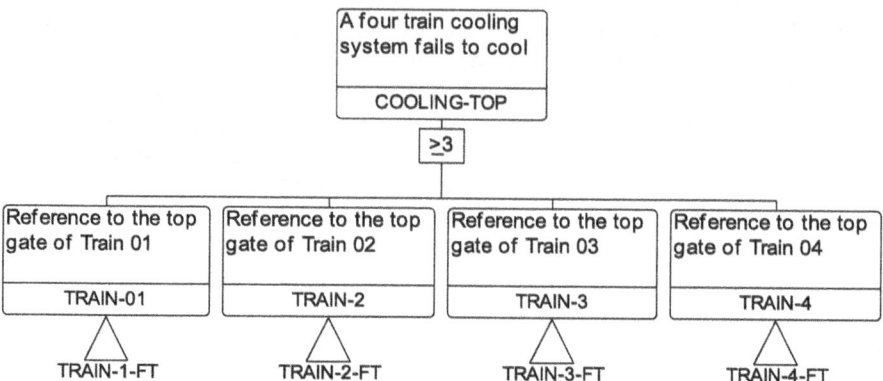

Fig. 6.2 The top structure of a fault tree for the cooling system with four redundant trains

the success criteria were the same then the fault tree top structure would also be the same: a 3-out-of-4 gate with the gates representing each of the places as inputs. Figure 6.2 depicts the top fault tree structure for this example system.

Let us look at the Train 1 in the example from Fig. 6.1. This train consists of three components in a series, a motor driven pump, a motor operated valve, and a check valve. The flow from the train is lost when the last component—the check valve—fails to function correctly or when there is no flow on the input of the last component. Assuming perfect pipes here, the flow on the input to the check valve is the same as the flow from the previous component, i.e., the motor operated valve. Recursively, there is no flow from the motor operated valve if it fails to function correctly or if there is no flow on its input. In one more step, we process the motor pump in the same way. Finally, we ask what can cause no flow on the input of the motor pump and in this simple case, there is neither a further component such as a water tank nor a link to another system. Therefore, we stop the recursive process here. The resulting fault tree is shown in Fig. 6.3.

In the next step, we develop fault trees for individual components. Let us look at the motor operated valve as an example. First, the expected functionality of this component is that it opens and stays open. The relevant failure modes will be 'fail to open' and 'spurious closure'. But the first failure mode might not always be relevant since the valve could already be open before we require the system to start. This system might be used in different scenarios and plant configurations. In some of them, the valve will be initially open, in some of them it will be initially closed. At the same time, we do not want to update the fault tree model when we start analyzing a different scenario. We need a possibility to turn this failure mode on or off at a later stage, when we define the analysis scenario.

House events—a special type of events that can be set to True or False depending on the plant configuration or the accident progression—are typically used as a standard modeling feature in these situations. We join the basic event for the failure mode 'fail to open' with a house event 'SYS_01_MV initially closed' under an AND-gate. When we want to consider this failure mode, we set this house event to True in the

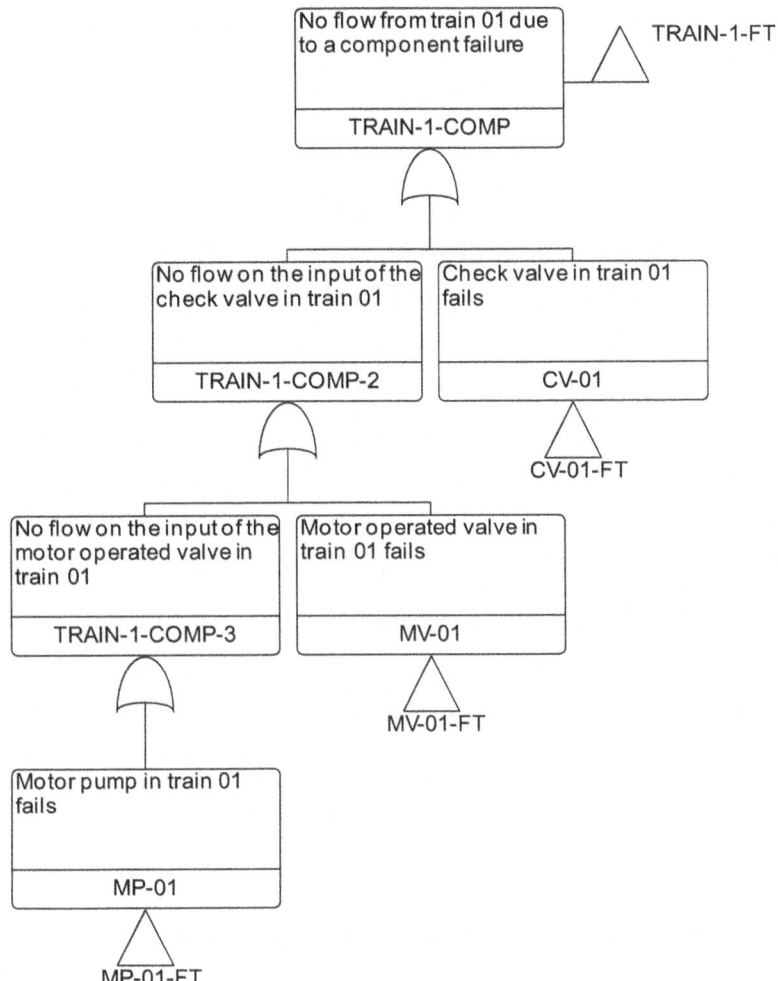

Fig. 6.3 A fault tree for the unavailability of Train 1 caused by unavailability of one of the components

definition of the analysis. Otherwise, when it is set to False, the failure mode will never occur in the accident scenarios.

In general, the same safety system can serve different purposes or at least it can achieve the same purpose in different ways, depending on the current plant configuration and the accident progression so far. The system fault tree model can contain all these purposes and ways in a single fault tree. This can be practical because it minimizes or eliminates duplication of the subsystem models. Failure logic relevant for a certain situation is selected with the help of house events.

Additionally, the correct functioning of this motor-operated valve depends on other systems such as power supply and signals. We assume that the FMEA identified these

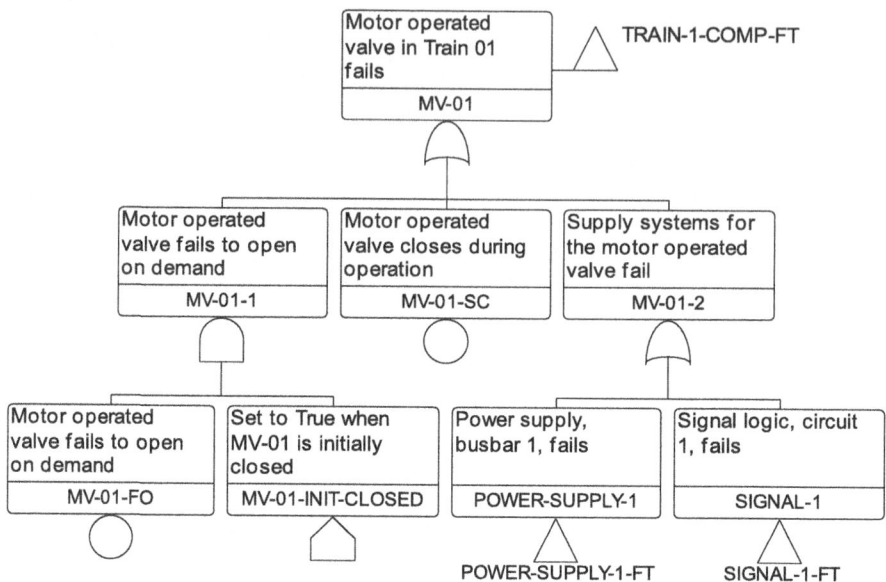

Fig. 6.4 A fault tree for the unavailability of the motor-operated valve in Train 1

two systems as the only ones relevant for the reliability analysis. Figure 6.4 displays a fault tree for the motor operated valve in Train 1 with a house event marking the initial position of the valve and with the two support systems modeled by transfers to their fault tree models. The recursive process continues there with the top event modeling the failure of the support system.

6.2.4 Component Groups: a Maintenance Example

Certain types of basic events correspond to a 'summary' failure somewhere in a block of components. This can be a pipe rupture anywhere in a train of serially connected components. It could be an operator failure that leads to the unavailability of the whole train. It could also be a take out of all components in a train due to preventive maintenance.

Let us explore modeling of preventive maintenance. We can define a new basic event for each train modeling unavailability of the whole train. Some safety-relevant information is not directly present in the design scheme. It could be information from the operating logic, such as that it is always the complete train taken out of operation for preventive maintenance. Let us look at consequences of this information for modeling.

First, let us explore a modeling alternative where the maintenance failure modes are listed as basic events in fault trees of individual components. In some cases, this gives us the most natural model. Assume that we would adopt this modeling strategy for Train 1 of the cooling system. Each component would have an additional basic event 'XX-01-MA' under the component top gate, where 'XX' stands for the component code. For example, the fault tree from Fig. 6.4 would have the basic event 'MV-01-MA' under the gate 'MV-01'. Now, each of the components can be unavailable because of maintenance. Let us see what happens if we update the fault tree in this way.

There are three different scenarios for the failure of Train 1—one where the motor pump is under maintenance, another one where the motor operated valve is under maintenance and yet another one where the check valve is under maintenance. This means that the unavailability of Train 1 because of maintenance is approximately three times higher than the unavailability due to maintenance of each of the components. In reality, they are not three different scenarios. In this plant, for practical reasons, preventive maintenance is performed on all components from the same train at the same time. Therefore, this is the same event and we would be triple-counting the same maintenance occasion. If the basic event 'MP-01-MA' occurs (the motor operated pump is under maintenance) then the basic event 'MV-01-MA' should occur as well with probability 1.0. This breaks the independence assumption for basic events. We either have to take this into account when analyzing the fault tree or we have to change the modeling.

From the reliability perspective, this model is equivalent to replacing all the maintenance basic events within one train by a single maintenance event which then occurs in the fault tree for every component. If there is maintenance work ongoing, all components in one train are unavailable. This, in turn, is equivalent to skipping the maintenance failure mode for individual components and treating it as a failure mode of the complete train. In the fault tree, we would then create a new top OR-gate with two inputs for each train. For Train 1, this gate will have the fault tree for no flow coming from Train 1 and the newly created maintenance basic event as inputs. Figure 6.5 depicts this OR-gate with the component failures and the maintenance as inputs.

Another example of safety relevant information not present in the design scheme is that at most one of the redundant trains will be taken out of operation for preventive maintenance at the same time. Therefore, failure scenarios where several trains are unavailable due to maintenance simultaneously cannot occur in reality. Our fault tree model contains these scenarios. We can either define several fault trees for each case of maintenance or we can extend our model with the information about mutual exclusivity of maintenance activities on different trains. This can be conveniently specified by using negated gates (NAND, NOR) or negated basic events (component A is unavailable because of maintenance AND it is not the case that component B is unavailable because of maintenance). Note that the mutual exclusivity might also affect the quantitative aspect of the analysis.

There could be other types of dependencies between components such as common cause failures. Failures of symmetrical pumps in four redundant trains are not completely independent. Failures can occur in several pumps during one mission, where these failures have the same underlying cause, for example that the same

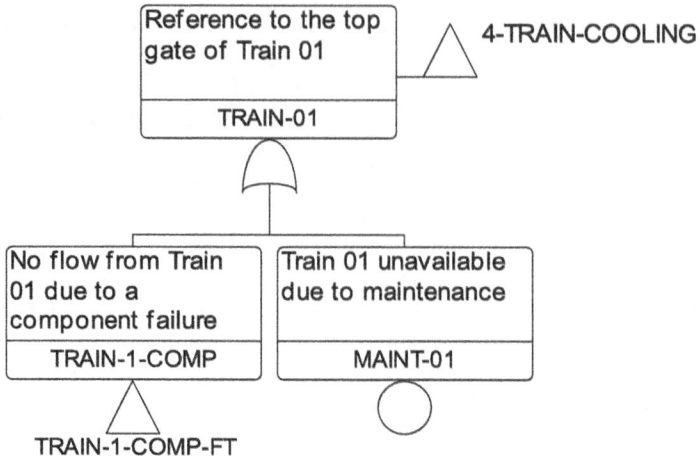

Fig. 6.5 A fault tree for the unavailability of Train 1 including the maintenance possibility

faulty lubricant was used during the last preventive maintenance occasion. We return to common cause failures in detail in Chap. 14.

Additional system failures might be related to already identified failure modes, but require several component failures to occur simultaneously. As an example, consider unintended leakage of the fluid from the system. This can happen, for instance, by a combination of inadvertently open valves. Another example might be a failure of a check valve that can lead to a backflow where a functioning pump does not transport the fluid to the sink, but it moves it in a loop within the pumping system. A good understanding of the plant under study and the dependability aspects are essential for the quality and completeness of the fault tree structure.

6.3 Model-Based Safety Assessment

The procedure described in Sect. 6.2.3 turns the safety relevant part of a design description into fault trees in a systematic way. Performing these steps manually requires skilled analysts and can lead to errors. Each analyst might prefer a different way of structuring fault trees. This results in visually different fault trees for symmetrical or very similar parts of the design. At the same time, the process of transforming a design into fault trees can be described by a possibly large yet well-formalized set of rules.

This sparked interest in tool support for creating fault trees and ultimately for building RAMS models on a high level, conceptually close to the language of system engineers. Fault trees can be automatically, in a rule-based manner, generated from high-level system descriptions by a computer program. Afterward, they can be analyzed by standard fault tree analysis tools, and the results can be interpreted back at the high-level.

This approach brings numerous advantages: time-saving when building large fault trees, minimizing manual errors, and consistency of the generated fault trees. It makes it easier to communicate with system engineers, keeping the link between the system-engineering model and the RAMS model. Defining a RAMS model at a high-level by connecting and configuring components and letting a tool to generate fault trees corresponds to writing a computer program in a high-level programming language and letting a compiler to produce the correct machine code. In this parallel, fault trees become a low-level, assembly-like, formalization of system or plant RAMS structure.

There are several established Model-Based Safety Assessment (MBSA) frameworks. Architecture Analysis and Design Language (AADL) [2] extended with Error Model Annex [3,4] allows architecture fault modeling and automated safety analysis by annotating AADL models with safety relevant information. There are tools for, e.g., fault tree analysis, failure mode and effects analysis, and Markov analysis of such annotated models. AltaRica [5,6] supports hierarchical modeling of components and their interaction. There is a tool support that automatically translates these high-level models to Guarded Transition Systems that can be further analyzed by software tools.

The modeling language Figaro [7] allows specifying components as classes, together with propagating effects of events (including failures) through the system. This description can serve either to automatically generate fault trees or, for systems with dynamic behavior, to translate them to Markov Decision Processes and analyze them by Monte Carlo simulations. Formal Safety Analysis Platform (FSAP) [8] performs safety analysis of a high-level system model by means of symbolic model checking. xSAP [9], a successor of FSAP, adds general modeling libraries and several safety analysis tools. In another framework called Hierarchically Performed Hazard Origin and Propagation Studies (Hip-HOPS) [10], individual components can be equipped by failure modes and mechanisms of their propagation. This allows for standard Fault Tree Analysis [11] and Failure Modes and Effects Analysis. Additionally, this formalism offers dynamic analyses, e.g., based on Petri nets [12].

6.3.1 Modeling Language Figaro

Here we exemplify the MBSA approach on the modeling language Figaro. It is an object-oriented language with declarative elements. Final classes correspond to system building blocks. For thermo-hydraulic systems, these can be standard components such as pumps, valves, heat exchangers, links connecting these components, or represent elements relevant for a safety analysis such as maintenance policy or operator failures. Each final class describes failure modes of the represented element and the failure propagation logic. This is done by two important constructs: interfaces and interaction rules. Interfaces encode the system structure. For each type of component relation, one can define a separate interface. For example, we can have an interface for topological relations such as all upstream or downstream neighbor components, we can encode dependencies such as power supply, cooling, or signals. More abstract relations, e.g., belonging to the same maintenance plan can be also captured by interfaces.

Interaction rules tell us how state changes in other components, possibly those related via interfaces, affect the state of this component or related components. One can specify sets of components by constructs with first-order logic quantifiers (is there an upstream component such that..., is ...true for all downstream components, ...). This gives us flexibility in the system structure specification. There is no need to declare how many components will be linked via a certain interface. Interaction rules are executed repeatedly until a fixed-point is reached. Internal state of components can be conveniently encoded by constants and variables.

Once we have specified components or system elements in a so-called *knowledge base* written in Figaro and validated it, a system engineer can employ it to model the analyzed system. The biggest part of the reliability engineering knowledge is now encapsulated in the component definitions. The analyst has to link all components correctly and configure the system for success criteria that shall be evaluated. A fully automated process traverses the system model and evaluates the interaction rules of the knowledge base to answer the question: 'What has to happen in the immediate upstream component (or a block of components) so that the unwanted failure at this place results from it.'

References

1. International Atomic Energy Agency (2024) Development and application of level 1 probabilistic safety assessment for nuclear power plants. No. SSG-3 (Rev. 1) in Specific Safety Guides, IAEA, Vienna
2. Society of Automotive Engineers (2017) AS5506C: architecture analysis and design language
3. Feiler PH, Gluch DP, Hudak J (2006) The architecture analysis and design language (AADL): an introduction. Carnegie Mellon University, Software Engineering Institute, Tech. rep
4. Delange J, Feiler PH (2014) Architecture fault modeling with the AADL error-model annex. In: 2014 40th EUROMICRO conference on software engineering and advanced applications, pp 361–368. https://api.semanticscholar.org/CorpusID:14117269
5. Point G, Rauzy AB (1999) Altarica: constraint automata as a description language. J Eur Syst Autom 33(8–9):1033–1052
6. Arnold A, Point G, Griffault A, Rauzy A (1999) The altarica formalism for describing concurrent systems. Fundamenta Informaticae 40(2, 3):109 124
7. Bouissou M, Bouhadana H, Bannelier M, Villatte N (1991b) Knowledge modelling and reliability processing: presentation of the figaro language and associated tools. In: IFAC symposium on safety of computer control systems 1991 (SAFECOMP'91), pp 69–75
8. Bozzano M, Villafiorita A (2007) The fsap/nusmv-sa safety analysis platform. Int J Softw Tools Technol Transf 9(1):5–24
9. Bittner B, Bozzano M, Cavada R, Cimatti A, Gario M, Griggio A, Mattarei C, Micheli A, Zampedri G (2016) The xSAP safety analysis platform. In: Proceedings of TACAS 2016

10. Papadopoulos Y, McDermid JA (1999) Hierarchically performed hazard origin and propagation studies. In: Felici M, Kanoun K (eds) Computer Safety, Reliability and Security, Springer Berlin Heidelberg, pp 139–152, https://doi.org/10.1007/3-540-48249-0_13
11. Papadopoulos Y, Maruhn M (2001) Model-based synthesis of fault trees from matlab-simulink models. In: Proceedings of international conference on dependable systems and networks (DSN 01), pp 77–82
12. Kabir S, Walker M, Papadopoulos Y (2018) Dynamic system safety analysis in hip-hops with petri nets and Bayesian networks. Saf Sci 105:55–70

Part III
Cut Sets

Minimal Cut Sets

7

Cut set analysis is one of the most fundamental analysis techniques for fault trees. A cut set is a set of basic events that makes the fault tree fail. Cut sets offer an analyst a set of failure scenarios that pinpoint critical failure paths. These are the foundation for effective risk management and decision-making, enabling vulnerability analysis and prioritization of critical failure scenarios.

In this chapter, we present the concept of cut sets, the dual concept of path sets, and their role in risk analysis. The next chapter (Chap. 8) presents quantification methods for cut sets and Chap. 9 discusses methods to compute all minimal cut sets from a fault tree.

7.1 Cut Sets

Systems are commonly designed to eliminate single points of failure—the phone in the road trip example being a typical instance. As a single component failure does not result in a system-level failure, multiple failures are required for the system to fail. Cut sets [1,2] indicate which failure combinations induce a system-level failure. That is, a cut set is a set of basic events that make the fault tree fail.

A *minimal cut set* is a cut set that cannot be reduced, i.e., no elements can be removed without losing its status as a cut set [3,4]. Note that a minimal cut set does not mean that its number of elements is minimal, but rather that no elements can be removed. Thus, a minimal cut set is a smallest combination of failures causing the top event to occur and the top event occurs if at least one of the minimal cut sets occurs.

© The Author(s), under exclusive license to Springer Nature Switzerland AG 2026
M. Stoelinga et al., *Concise Guide to Fault Tree Analysis*, Computer Science Foundations and Applied Logic, https://doi.org/10.1007/978-3-031-78287-9_7

Definition 10 A *cut set* is a set of basic events that together cause the top event to occur. A *minimal cut set* is a cut set of which no proper subset is also a cut set.

Figure 7.1 presents some cut sets for the road trip example. From now on, our road trip example uses abbreviated event names, listed in Table 7.1 on page 80. Mathematically, a cut set of a fault tree F is a set of basic events whose structure function returns 1, that is, a set $C \subseteq BE_F$ such that $\Phi_F(C) = 1$. A cut set C is minimal if no proper subset of C is a cut set, i.e., for all $C' \subset C$, we have $\Phi_F(C') = 0$.

7.2 Coherency, Absence of Order, and Probabilistic Independence

Coherency. As mentioned in Sect. 2.4.2, this book focuses on fault trees with AND, OR, and voting gates. These fault trees are *coherent*, meaning that a cut set remains a cut set after adding elements to it. Mathematically: for all cut sets $C \subseteq BE_F$ and events $c \in BE_F$, if C is a cut set of F, then $C \cup \{c\}$ is also a cut set of F.

It is important to realize that, due to coherency, a cut set does not encode a single failure scenario where all elements in the cut set have failed, but other basic events have not. *Rather, a cut set represents a set of scenarios where the elements in the set have failed and the status of others is unknown, i.e., they can be either failed or operational.*

Fault trees with NOT-based gates, such as the exclusive OR (XOR) gate and NAND gate, are usually not coherent: For example, if $F = \mathsf{XOR}(e, e')$, then $\{e\}$ is a cut set, but $\{e, e'\}$ is not a cut set. Non-coherent extensions require different analysis methods, discussed in Chap. 15.

Absence of temporal order. Due to the static nature of standard fault trees, a cut set does not contain any information about incident progression or the temporal order of failures. An experienced analyst or system engineer may be able to imagine a system behavior corresponding to a particular cut set, but from a mathematical perspective, a cut set is just a set. There is no sequential order of events expressing which failure occurred first and which followed. Neither is any information specified about how much time has passed between the first and the last event. Cut sets represent a timeless snapshot of the modeled system.

Dynamic fault trees do consider time-dependent behavior. As a result, cut sets no longer suffice to describe the failure behavior. Instead, failure sequences are used [5]. Since dynamic fault trees are also non-coherent, such failure sequences do require additional care [6]. Section 16.1.1 provides more details.

Probabilistic independence. Importantly, Definition 13 assumes that all basic events are *probabilistically independent*. That is, information about the failure of one event does not increase or decrease the failure probabilities of other events. If the basic events are not stochastically independent, then methods for common cause failures should be considered, as described in Chap. 14.

Example 10 (Cut sets of the road trip example)

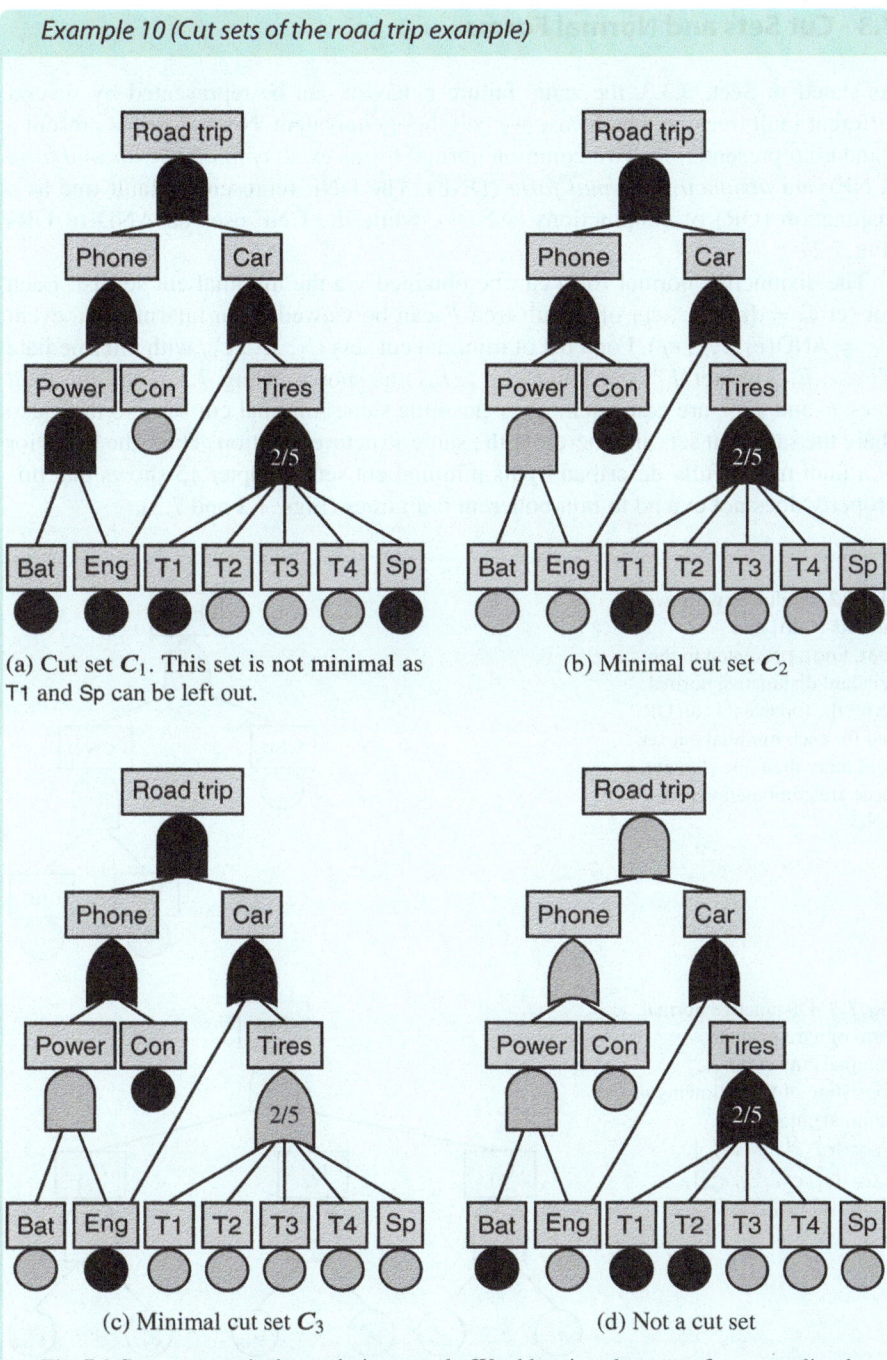

(a) Cut set C_1. This set is not minimal as T1 and Sp can be left out.

(b) Minimal cut set C_2

(c) Minimal cut set C_3

(d) Not a cut set

Fig. 7.1 Some cut sets in the road trip example. We abbreviate the names for events, listed in Table 7.1 on page 80. Black basic events indicate elements of the cut sets. The black gates show how these failures propagate through the system. The full list of cut sets is given in Table 7.2

7.3   Cut Sets and Normal Forms

As stated in Sect. 2.3.3, the same failure behavior can be represented by several different fault trees, in which case we call these equivalent. Normal forms present a standard representation. Two common normal forms exist: *conjunctive normal form* (CNF) and *disjunctive normal form* (DNF). The DNF represents a fault tree as a disjunction (OR) of conjunctions (ANDs), while the CNF uses an AND of ORs (Fig. 7.2).

The disjunctive normal form can be obtained via the minimal cut set list. Each cut set $C = \{e_1, \ldots, e_k\}$ of a fault tree F can be viewed as an intermediate event $E_C = \mathsf{AND}(e_1, \ldots, e_k)$. For a list of minimal cut sets $C_1, \ldots, C_m$ with intermediate $E_1 \ldots, E_m$, we set $F^{\mathrm{dnf}} = \mathsf{OR}(E_1, \ldots, E_m)$ as shown in Fig. 7.3. Then the fault trees F and F^{dnf} are equivalent: they have the same minimal cut sets, so they also share the same cut sets and therefore the same structure function. Thus, the behavior of a fault tree is fully described by its minimal cut sets. Chapter 15 shows that this property does not extend to non-coherent fault trees (Figs. 7.2 and 7.3).

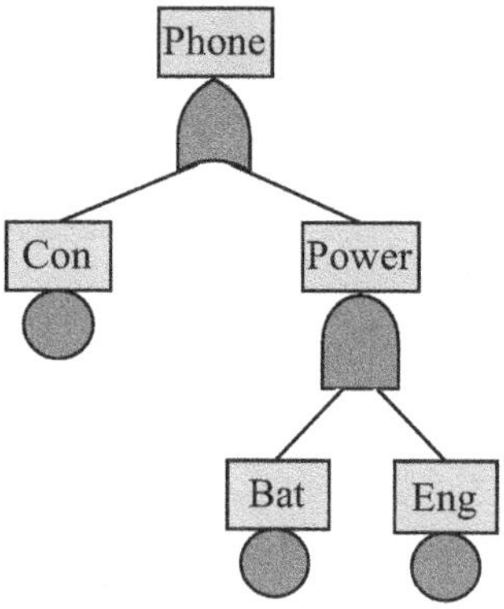

Fig. 7.2 Fault tree with two cut sets {Con} and {Bat, Eng}, presented in the standard disjunctive normal form: the top event is an OR, and for each minimal cut set with more than one element, these are combined via an AND

Fig. 7.3 Disjunctive normal form of a tree via the minimal cut set list L, consisting of the following m minimal cut sets:
$C_1 = \{e_1^1, e_2^1, \ldots, e_{n_1}^1\}$,
$C_2 = \{e_1^2, e_2^2, \ldots, e_{n_2}^2\}$, ...,
$C_m = \{e_1^m, e_2^m, \ldots, e_{n_m}^m\}$

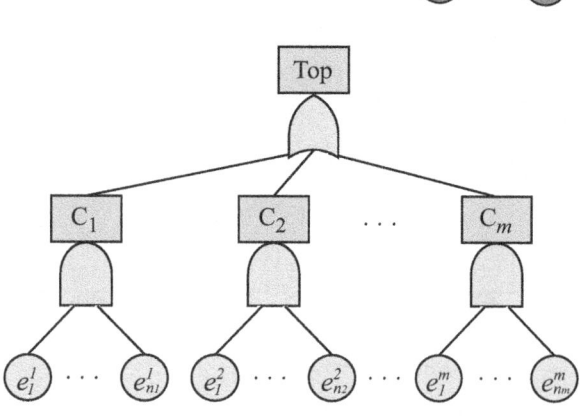

7.4 Cut Set Metrics

Metrics quantify the importance of a cut set. We discuss three common metrics for cut sets: the order, frequency, and probability. These metrics make the most sense for minimal cut sets, but their definitions apply to any cut set.

Definition 11 *The order* of a cut set is the number of elements contained in this set.

The cut set order is commonly used to identify system vulnerabilities and error-prone situations: Cut sets of low order indicate a vulnerability, especially if their elements are likely to fail or there are uncertainties about this likelihood.

Definition 12 *The frequency* of the basic event e in a family of minimal cut sets is the number of minimal cut sets containing e.

Apart from the order of a cut set, the *frequency* of an element in a cut set is important. If the same element appears in many cut sets (especially of low order), this also indicates a vulnerability.

Cut set probabilities are an important tool in quantitative analysis, as they quantify the likelihood for a cut set to fail. To do so, probabilistic information is required for all basic events.

Definition 13 If each basic element e_i is assigned a probability p_i to fail, then the *cut set probability* $\mathbb{P}[C]$ for a cut set $C = \{e_1, \dots, e_m\}$ to fail is given by $\mathbb{P}[C] = p_1 \cdot \dots \cdot p_m$.

Thus, $\mathbb{P}[C]$ computes the probability that all elements in C fail. Elements outside C may or may not fail. Defining $\mathbb{P}[C] = p_1 \cdot \dots \cdot p_m$ crucially relies on the stochastic independence of the basic events, see Sect. 14.1. Note that this notation differs from other notation in probability theory, where for a set A, the probability P[A] denotes that some elements in A occur. E.g., for a dice, P[{2,3}] means that either outcome 2 or 3 comes up. However, when we write P[{2,3}], then we mean that both events 2 and 3 have to happen.

Table 7.1 Properties of basic events

Basic event	Full name	Probability	Frequency
Con	No connection	0.25	11
Bat	Battery fails	0.2	1
Eng	Engine fails	0.05	2
Ti	Tire i fails	0.1	4
Sp	Spare fails	0.1	4

Table 7.2 Metrics for cut sets: order and probability

Minimal cut set	Order	Probability	
Con, Eng	2	$0.25 \cdot 0.05$	= 1.25E-2
Bat, Eng	2	$0.2 \cdot 0.05$	= 1.0E-2
Con, T1, T2	3	$0.25 \cdot 0.1 \cdot 0.1$	= 2.5E-3
Con, T1, T3	3	$0.25 \cdot 0.1 \cdot 0.1$	= 2.5E-3
Con, T1, T4	3	$0.25 \cdot 0.1 \cdot 0.1$	= 2.5E-3
Con, T1, Sp	3	$0.25 \cdot 0.1 \cdot 0.1$	= 2.5E-3
Con, T2, T3	3	$0.25 \cdot 0.1 \cdot 0.1$	= 2.5E-3
Con, T2, T4	3	$0.25 \cdot 0.1 \cdot 0.1$	= 2.5E-3
Con, T2, Sp	3	$0.25 \cdot 0.1 \cdot 0.1$	= 2.5E-3
Con, T3, T4	3	$0.25 \cdot 0.1 \cdot 0.1$	= 2.5E-3
Con, T3, Sp	3	$0.25 \cdot 0.1 \cdot 0.1$	= 2.5E-3
Con, T4, Sp	3	$0.25 \cdot 0.1 \cdot 0.1$	= 2.5E-3
TOTAL			= 4.75E-2

Example 11 (Cut set metrics)

Tables 7.1 and 7.2 illustrate the cut set metrics for the road trip example.

- *Frequencies.* Table 7.1 lists frequencies for all basic events.
- *Orders.* Table 7.2 lists all minimal cut sets and the order of each one.
- *Probabilities.* Table 7.2 gives the cut set probabilities, assuming basic event failure probabilities as in Table 7.1. To make the computation steps more tractable, we have taken unrealistically high failure probabilities for the basic events.

7.5   Cut Set Analysis

Cut set analysis is a cornerstone method in fault tree analysis, initiating efficient risk management strategies and informed decision-making processes. As before, we distinguish between qualitative and quantitative analysis. Qualitative cut set analysis examines cut sets to validate fault trees and to identify common cause failures and vulnerabilities. Quantitative analysis uses cut set probabilities to approximate system failure probabilities, prioritize critical cut sets for risk management, and demonstrate compliance with dependability requirements.

Cut set analysis requires a list of minimal cut sets to be available. Computing this list can be complex for large fault trees and is the topic of the next chapter.

7.5.1   Qualitative Analysis

Validation of the fault tree. In practical systems, fault trees can become very large, increasing the likelihood of errors in the modeling process. By validating fault tree models, their alignment with real-world observations is confirmed, enhancing trust in their ability to describe and predict outcomes. One way to do so is to check the reasonableness of all minimal cut sets, or at least a large number of them: Is it indeed the case that the failures of these basic events make the system fail, or are additional failures required? Are the generated minimal cut sets indeed minimal, or can we leave out events? In addition, path set analysis (cf. Sect. 7.7) can reveal overlooked failure scenarios.

Identification of system vulnerabilities. A primary aim of cut set analysis is to pinpoint system vulnerabilities, so that effective measures can be taken to eliminate, reduce, or mitigate these. To do so, one examines cut sets of low order, as well as elements with high cut set frequencies.

A cut set with just a few elements, or with elements whose failure is considered likely, indicates a system vulnerability. Therefore, an important task is to assess whether (low-order) cut sets give rise to design improvements. In addition, the *frequency* of an element in the list of all cut sets is important. If the same element appears in many (low-order) cut sets, this also indicates a vulnerability and may give rise to design improvements.

Moreover, if the estimated likelihood of event occurrences is reasonably low, then the uncertainties in the likelihood estimate will play a much larger role in low-order cut sets. Additionally, more events in a cut set typically mean more barriers or mitigating systems that give us more time to react to the progression of the accident and apply measures not included in the fault tree.

The cut set $C_3 = \{$No connection, Engine fails$\}$ in Example 7.1 is typical here. With only two elements, it has a very low order. Moreover, one of its elements, namely No connection, is likely to occur: connection failures happen regularly, especially in remote areas. Hence, this cut set reveals a vulnerability; a design improvement can be to bring a separate satellite telephone, thereby adding redundancy as well as adding a lower-probability component to the minimal cut set.

Fig. 7.4 Ranking the cut set probabilities

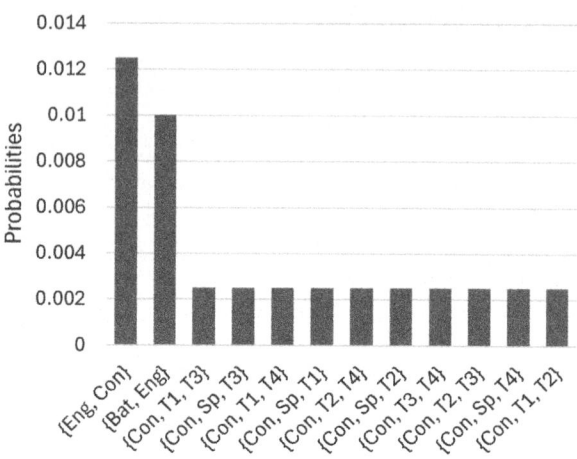

Identification of common cause failures. Common cause failures are factors that cause failures in multiple components at the same time, for instance due to a fire, manufacturing defect, or harsh environmental conditions. Common causes often dominate the system dependability and are therefore an important concern for risk analysis. Inspecting the cut sets in the road trip example reveals that the tires are subject to common cause failures, for instance, due to production errors or bad road conditions.

Checking minimal cut sets for independence is of crucial importance. If a minimal cut set contains basic events that are subject to common causes, then these should be properly treated. There are several ways to accomplish this task, discussed in Chap. 14.

7.5.2   Quantitative Analysis

Ranking minimal cut sets. An advantage of computing cut set probabilities is that we can see the relative importance of cut sets: Figure 7.4 displays the cut set probabilities from Table 7.2 and clearly shows that the first two cut sets dominate. Hence, investing in decreasing the probabilities for these cut sets has the greatest impact on the system reliability. Another application of this quantitative method is the ability to evaluate the impact of interventions. For example, how would the cut set probabilities be affected by the introduction of a satellite phone, where the connection has a failure probability of 0.08 instead of 0.25?

Approximation of the top event probability. One of the key quantitative measures for a fault tree is p_{Top}, the probability of the top event to happen. Section 10.2 provides several calculation methods for obtaining the exact value of the top event probability from the basic event probabilities. However, if the fault tree is large and/or contains many shared events, the exact probability can be difficult or even infeasible to calculate.

In that case, cut sets provide an effective way to approximate the probability p_{Top} by adding up all cut set probabilities. This is a conservative approximation: the true top probability p_{Top} is always lower than the approximation via the sum of minimal cut sets. Therefore, the sum-of-minimal-cut-sets approximation $p_{\approx}$ is safe. That is, if the approximation $p_{\approx}$ is below the maximally tolerated failure probability $p_{\max}$ then the exact probability p_{Top} is also below $p_{\max}$, since $p_{Top} \leq p_{\approx} \leq p_{\max}$.

If $C_1, \ldots, C_k$ are all minimal cut sets in a fault tree, then two upper bound equations hold.

$$p_{Top} \leq \mathbb{P}[C_1] + \cdots + \mathbb{P}[C_k] \qquad \text{(REA)}$$
$$p_{Top} \leq 1 - \big((1 - \mathbb{P}[C_1])(1 - \mathbb{P}[C_2]) \cdots (1 - \mathbb{P}[C_k])\big) \qquad \text{(MCUB)}$$

These equations, called the *rare event approximation (REA)* and *min cut upper bound approximation (MCUB)* respectively, are further elaborated in Chap. 8.

7.5.3   Time-Dependent Probabilistic Cut Set Measures

Cut sets can also be used to study time-dependent failure probabilities. Time-dependent failure behavior is extensively discussed in Chap. 11. For now, we let $p_i(t)$ represent the (point) unavailability at time t. That is, $p_i(t)$ is the probability that e_i has failed at time t. Then the time-dependent cut set probability of cut set C is given by $\mathbb{P}[C](t) = p_1(t) \cdot \ldots \cdot p_m(t)$. The REA and MCUB approximations also extend to the time-dependent case:

$$p_{Top}(t) \leq \mathbb{P}[C_1](t) + \cdots + \mathbb{P}[C_k](t)$$
$$p_{Top}(t) \leq 1 - \big((1 - \mathbb{P}[C_1](t))(1 - \mathbb{P}[C_2](t)) \cdots (1 - \mathbb{P}[C_k](t))\big)$$

For example, if $p_{Bat}(t) = 1 - e^{-\lambda_1 t}$ and $p_{Con}(t) = 1 - e^{-\lambda_2 t}$, that is, Bat and Con have exponential failure time distributions, then the Cumulative Distribution Function (CDF) of the cut set failure time equals the product of the component CDFs, i.e., $\mathbb{P}[\{Bat, Con\}](t) = (1 - e^{-\lambda_1 t})(1 - e^{-\lambda_2 t})$.

7.6   The Process of Cut Set Analysis

Assuming a fault tree and relevant data about the cut set probabilities are available, the process of cut set analysis proceeds according to the following steps:

1. *Determine the list of minimal cut sets.* Clearly, cut set analysis starts with a list of all minimal cut sets, e.g., ranked with respect to their order.
2. *Validate these cut sets.* Check if these minimal cut sets make sense: do they indeed cause a system failure? Were basic events forgotten, or left out? Are some basic events unnecessary?

3. *Investigate common cause failures.* As stated, common cause failures often dominate the system's failure behavior. Common cause failures can be investigated by examining whether multiple basic events in a minimal cut set are subject to a common cause. If so, one can adjust the cut set probability or the fault tree, as explained in Chap. 14.

4. *Compute cut set ranking.* Rank the cut sets based on relevant metrics to identify those with the highest contribution to system risk.

5. *Find mitigating measures for the cut sets with the highest rank.* Once the cut sets with the highest risk are found, one devises measures to eliminate, mitigate, or reduce these risks.

6. *Add these measures to the fault tree model, and re-evaluate.* Once mitigation measures have been proposed, it is good practice to re-evaluate the system and see if these measures have helped. Adjusting the fault tree to reflect the mitigations and recomputing the cut set metrics is a good way to evaluate the effect of these measures.

7.7 Path Sets

Path sets are the dual of cut sets, indicating which combinations of basic events keep the system operational. Thus, path sets could be considered success sets. They serve the same purposes as cut sets, however, in practice cuts sets are used far more often.

Definition 14 A *path set* is a set of basic events such that, if they do not fail, then the system remains operational. A *minimal path set* is a path set of which no proper subset is a path set.

Mathematically, a *path set* of a fault tree F is a set $P \subseteq BE$ such that $\Phi_F(BE \backslash P) = 0$. In addition, P is a minimal path set if for all $P' \subset P$, we have $\Phi_F(BE \backslash P') = 1$.

Path sets serve the same purposes as cut sets in qualitative and quantitative analysis of fault trees. In particular, minimal path sets can be used for dependability improvements. Path set with just a few elements give confidence, as a few components keep the system operational. Large path sets point to vulnerabilities, requiring many components for the system to operate properly. Measures to reduce large path sets will increase the overall dependability.

Importantly, in validation, path sets can provide an additional perspective on the completeness of the fault tree: By focusing on successes, the examination of path sets can reveal whether no failure scenarios were forgotten. This is exemplified by the set {T1, T2, T3, Sp} that is a path set of the Tires subtree. One may question if this set is complete for the success of the tire system, since equipment and skills are essential to mount the spare tire on the car.

Cut sets versus path sets. There is a strong relation between cut sets and path sets: Each path set in a fault tree F is a cut set in a fault tree F' (called the success tree) that we obtain from F by inverting all gates: AND becomes OR; OR becomes AND and a k/N VOTING-gate becomes an $(N - k + 1)/N$ VOTING-gate. Further, we replace each basic event by its complement (i.e., 'component failure' by 'no component failure'). Thus, methods to compute cut sets also work for computing path sets, by running them on the transformed tree F'. These computation methods are discussed in the next chapter.

Example 12 (Path sets in the road trip example)

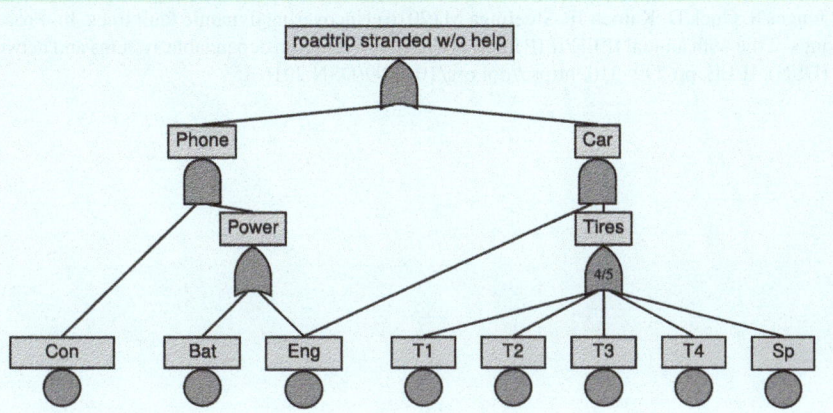

The figure above displays the success tree of the road trip example: the diagram is the same as before, but with inverted gates: AND becomes OR; OR becomes AND and the 2/5 voting-gate is now a 4/5 voting-gate. The success tree above has, among others, the following cut sets, which are the path sets of the fault tree in Fig. 2.1.

- $P_1 = \{\text{Bat}, \text{Con}\}$: If the phone battery is not empty and there is a connection, then the road trip does not strand, because one can call road services. Note that, indeed, $\Phi(BE \setminus P_1) = \Phi(\{\text{Eng}, \text{T1}, \text{T2}, \text{T3}, \text{T4}, \text{Sp}\}) = 0$, so P_1 fulfills the condition in Definition 14 for being a path set.
- $P_2 = \{\text{T1}, \text{T3}, \text{T4}, \text{Sp}, \text{Eng}\}$: If 4 tires and the engine are operational, then one can drive the car, hence the road trip has not failed. One may actually question if P_2 should really be a path set. For example, is no equipment needed to mount the spare on the car after Tire 2 has failed? Posing such critical questions for validation of the fault tree is a valuable activity.

References

1. Vatn J (1992) Finding minimal cut sets in a fault tree. Reliab Eng Syst Saf 36(1):59–62
2. Rauzy A (2001) Mathematical foundations of minimal cutsets. IEEE Trans Reliab 50(4):389–396
3. Vesely WE, Goldberg FF, Roberts NH, Haasl DF (1981) Fault tree handbook. Nuclear Regulatory Commission, Office of Nuclear Regulatory Research, U.S
4. Fussell JB, Vesely WE (1972) New methodology for obtaining cut sets for fault trees. Transactions of the American Nuclear Society
5. Tang Z, Dugan JB (2004a) Minimal cut set/sequence generation for dynamic fault trees. In: Annual symposium reliability and maintainability, 2004-RAMS, IEEE, pp 207–213
6. Junges S, Guck D, Katoen JP, Stoelinga M (2016) Uncovering dynamic fault trees. In: Proceedings of the 46th annual IEEE/IFIP international conference on dependable systems and networks (DSN), IEEE, pp 299–310. https://doi.org/10.1109/DSN.2016.35

Probabilistic Analysis via Minimal Cut Sets

Calculation of the top event probability plays a central role in quantitative fault tree analysis, as it serves as a basis for many other dependability metrics, such as the availability, reliability, and component importance. This value can be computed exactly using the techniques from the next chapter. However, for large fault trees, the exact calculation can be infeasible. In such cases, the top event probability can be efficiently approximated using the minimal cut set list, calculating the probability of each cut set. This chapter outlines three of these methods, detailing their assumptions and precision.

Before delving into the mathematical and algorithmic details of estimating top event probabilities, it is crucial to address the uncertainties in quantifying basic failures. Even with a deep understanding of a component's operation and failure modes, we can only create a mathematical model of its basic failures. For static fault tree analysis, we need a simple model that provides a failure probability. Given the inherent uncertainties, abstractions, and simplifications involved, the best we can achieve is a reasonably accurate approximation.

For example, if the aim is to identify design improvements to meet tighter availability requirements, an overly conservative estimate might incorrectly flag a component for redundancy. In reality, a different part of the system may need duplication. Recognizing these uncertainties can shift how we view efforts to calculate precise top event probabilities. What is the value of an exact sum of imprecise numbers? Sometimes, approximations may have a smaller effect than uncertainties, still allowing us to use them effectively. The goal is not mathematical precision, but practical insights for decision-making based on risks of an analyzed system.

This section provides three methods to conservatively approximate the top event probability based on cut set probabilities.

M. Stoelinga et al., *Concise Guide to Fault Tree Analysis*, Computer Science Foundations and Applied Logic, https://doi.org/10.1007/978-3-031-78287-9_8

- The *rare event approximation* approximates the top probability via the sum of all minimal cut set probabilities.
- The *min cut upper bound* provides a tighter bound and is based on negating cut set probabilities.
- *Higher order approximations* extend these approximations by considering combinations of minimal cut sets. They provide even tighter bounds on the probability, but also require more computational effort.

Definition 15 The top event probability $\mathbb{P}[F]$ for a fault tree F with the structure function Φ_F is defined as follows.

- The probability $\mathbb{P}[b_i]$ of a single status bit b_i is given by $\mathbb{P}[b_i] = \mathbb{P}[e_i]$ if $b_i = 1$ (i.e., e_i has failed) and $\mathbb{P}[b_i] = 1 - \mathbb{P}[e_i]$ if $b_i = 0$ (i.e., e_i has not failed). This can be written as

$$\mathbb{P}[b_i] = b_i \cdot \mathbb{P}[e_i] + (1 - b_i) \cdot (1 - \mathbb{P}[e_i])$$

- The probability of a status vector is obtained by multiplying the probabilities of all its status bits, i.e.,

$$\mathbb{P}[\mathbf{b}] = \mathbb{P}[b_1] \cdots \mathbb{P}[b_n] = \prod_{1 \leq i \leq n} b_i \cdot \mathbb{P}[e_i] + (1 - b_i) \cdot (1 - \mathbb{P}[e_i])$$

- Adding the probabilities of all status vectors on which the structure function fails (i.e., evaluates to one) yields the probability that the top event occurs:

$$\mathbb{P}[F] = \sum_{\mathbf{b} \in \{0,1\}^n} \Phi_F(\mathbf{b}) \cdot \mathbb{P}[\mathbf{b}]$$

8.1 Rare Event Approximation

Let us for a minimal cut set list L denote by $\mathrm{REA}(L)$ the *rare event approximation* of L, as the sum of probabilities of all cut sets from L:

$$\mathrm{REA}(L) = \sum_{C \in L} \mathbb{P}[C]$$

The approximation conservatively estimates the top event value. That is, if L_F denotes the list of cut sets of a fault tree F, then it is always the case that

$$\mathrm{REA}(L_F) \geq \mathbb{P}[F]$$

This statement is proven in the appendix of this chapter, starting on page 93.

Fig. 8.1 Illustration of the rare event approximation for three MCSs. $\mathbb{P}[C_1 \cup C_2 \cup C_3]$ is the area covered by any circle. The rare event approximation adds the areas of C_1, C_2 and C_3 individually, which is larger than their joint area.

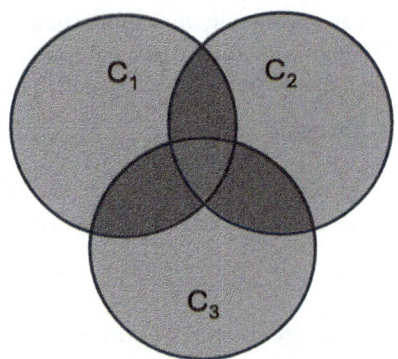

Conceptual illustration. The rare event approximation can be illustrated in a Venn diagram (Fig. 8.1): The probability of each cut set is visualized as the area of a circle. Then, the probability of all sets together, i.e., $\mathbb{P}[C_1 \cup C_2 \cup C_3]$, equals the area covered by all circles together. The rare event approximation, however, adds the areas of each individual circle. Clearly, this overapproximates the area jointly covered by the circles. In particular, the area of the intersection $C_1 \cap C_2$ is added twice; the area of $C_1 \cap C_2 \cap C_3$ is added three times. Recall that P[C] denotes the probability that the cut set C occurs, i.e., that all events in C occur. Thus, we have $P[C_1 \wedge C_2] = P[C_1 \cup C_2]$, which is not the case if C_1 and C_2 are interpreted as normal sets (where we have $P[C_1 \wedge C_2] = P[C_1 \cap C_2]$).

Example 13

Consider the fault tree $F = \mathrm{OR}(e_1, e_2)$. Its minimal cut set list L contains two minimal cut sets, namely $C_1 = \{e_1\}$ and $C_2 = \{e_2\}$. The top event occurs if one of these occurs. The exact top event value can be calculated by summing up the probabilities of status vectors that fail the top event, namely $10, 01, 11$:

$$\mathbb{P}[F] = \mathbb{P}[10] + \mathbb{P}[01] + \mathbb{P}[11]$$

Calculating the probability of each minimal cut set as a sum of the status vector probabilities yields

$$\mathbb{P}[\{e_1\}] = \mathbb{P}[10] + \mathbb{P}[11]$$
$$\mathbb{P}[\{e_2\}] = \mathbb{P}[01] + \mathbb{P}[11]$$

A sum of probabilities of minimal cut sets is then

$$\mathrm{REA}(L) = \mathbb{P}[\{e_1\}] + \mathbb{P}[\{e_2\}] = \mathbb{P}[10] + \mathbb{P}[11] + \mathbb{P}[01] + \mathbb{P}[11]$$

Indeed, $\mathbb{P}[F] + \mathbb{P}[11] = \mathrm{REA}(L)$, $\mathbb{P}[F] \leq \mathrm{REA}(L)$.

The reason why this approximation is called the rare event approximation is that this approximation works especially well for rare events. That is, the bounds are tight when the probabilities are small. This can be seen in a small example.

Example 14

Consider the fault tree $F = \text{OR}(e_1, e_2)$ again. We write $p_1 = \mathbb{P}[e_1]$ and $p_2 = \mathbb{P}[e_2]$. The exact top event probability is given by

$$\mathbb{P}[F] = \mathbb{P}[e_1 \vee e_2] = p_1 + p_2 - p_1 \cdot p_2$$

The rare event approximation is given by

$$\text{REA}(C_1, C_2) = p_1 + p_2$$

Compare three scenarios.

- Case 1: $p_1 = p_2 = 0.5$, then

$$\mathbb{P}[F] = 0.5 + 0.5 - 0.25 = 0.75 \leq \text{REA}(C_1, C_2) = 1$$

- Case 2: $p_1 = p_2 = 0.5 \cdot 10^{-2}$, then

$$\mathbb{P}[F] = 0.5 \cdot 10^{-2} + 0.5 \cdot 10^{-2} - 0.25 \cdot 10^{-4}$$
$$= 0.09975 \leq \text{REA}(C_1, C_2) = 0.1$$

- Case 3: $p_1 = p_2 = 0.5 \cdot 10^{-4}$, then

$$\mathbb{P}[F] = 0.5 \cdot 10^{-4} + 0.5 \cdot 10^{-4} - 0.25 \cdot 10^{-8}$$
$$= 9.99975 \cdot 10^{-4} \leq \text{REA}(C_1, C_2) = 1 \cdot 10^{-3}$$

In the first case, the rare event approximation adds 33% of the exact top probability.

This method safely over-estimates the top event value when minimal cut sets are independent of each other, and also when there are correlations between minimal cut sets, both positive and negative.

For the case of independent basic events, let us assume an arbitrary fault tree F and the corresponding list of minimal cut sets L. Probability of each minimal cut set $C \in L$ can be expressed as a sum of its satisfying status vectors. $\text{REA}(L)$ is then the sum of sums of status vectors for minimal cut sets in L. $\text{REA}(L)$ will always contain all status vectors that satisfy F, because for each status vector satisfying F, there is a minimal cut set that is also satisfied by this status vector. This follows from the fact that $\Phi(L) = \Phi(F)$ and the fact that cut sets in L are connected by OR. On the other hand, there might be status vectors in $\text{REA}(L)$ that are counted several times. For two minimal cut sets C_1, C_2, let us take a status vector for $C_1 \cup C_2$. Any such status vector will be calculated in REA(L) at least twice, because it is included in $\mathbb{P}[C_1]$ and in $\mathbb{P}[C_2]$.

Approximation. In many practical situations, rare event approximation produces results very close to the top event value. Fault trees model reliable systems where basic failures are typically rare. For each status vector that is counted at least twice, there is another status vector at least two orders of magnitude greater. Status vectors counted several times will then increase the top value at most by a percent each.

Using an approximate value requires considering its adequacy for the analysis purpose. For safety cases aiming to demonstrate that risks are below regulatory limits, a conservative approximation may be sufficient. Similarly, when informing design or operational decisions, an approximation within the range of data uncertainties can be acceptable, as increased precision might not impact the outcome. However, excessive conservatism can distort decision-making by misranking components or subsystems based on their actual risk contribution, potentially leading to inefficient allocation of resources.

8.2   Min Cut Upper Bound Approximation

Let us for a minimal cut set list L denote by MCUB(L) the numerical disjunction of probabilities of cut sets from L:

$$\text{MCUB}(L) = 1 - \prod_{C \in L} (1 - \mathbb{P}[C])$$

This approximation considers all minimal cut sets as independent events and quantifies their disjunction exactly (under this assumption). We can see it as an exact quantification of a fault tree with the top OR gate with basic events for each minimal cut set as inputs, where each basic event has the probability of the corresponding minimal cut set.

It can be shown that the min cut upper bound is always lower than the rare event approximation. Therefore, MCUB gives tighter, i.e., less conservative bounds on the top probability.

$$\mathbb{P}[F] \leq \text{MCUB}(L) \leq \text{REA}(L)$$

The proof of this property is given in Appendix of this chapter, starting on page 93.

Example 15

The example from the rare event approximation with L containing $C_1 = \{e_1\}$ and $C_2 = \{e_2\}$ will be quantified exactly by the min cut upper bound. The formula $\text{MCUB}(L) = 1 - (1 - \mathbb{P}[C_1]) \cdot (1 - \mathbb{P}[C_2])$ evaluates to $\mathbb{P}[C_1] + \mathbb{P}[C_2] - \mathbb{P}[C_1] \cdot \mathbb{P}[C_2]$. The product $\mathbb{P}[C_1] \cdot \mathbb{P}[C_2]$ is equal to $\mathbb{P}[C_1 \cup C_2] = \mathbb{P}[\{e_1, e_2\}]$, because C_1 and C_2 are independent. This means that this product is equal to the probability of the status vector 11 which is counted twice in the sum $\mathbb{P}[C_1] + \mathbb{P}[C_2]$, as we have seen above.

Approximation. The precision of MCUB depends to a big extent on the same factors as the rare event approximation. Min cut upper bound does not take dependencies between minimal cut sets into account either. On the other hand, it always gives a better estimate when basic events are independent.

8.3 Higher Order Approximations

There is a systematic way how to successively reduce the conservatism of the rare event approximation and eventually converge to the exact value of the top event probability. This method uses the so-called *inclusion-exclusion* principle. It is based on the observation that status vectors that are counted several times by the rare event approximation are those that fail a pair of minimal cut sets. Therefore, we take all pairs of minimal cut sets and for each of them create the union of the two minimal cut sets. This gives us a new cut set, even though not minimal. We subtract the probability of this cut set from the rare event approximation. By this, we remove (exclude) all status vectors that fail both minimal cut sets at the same time. Because we look at pairs of minimal cut sets, this approximation is called the *second order approximation*, denoted APPROX-2.

$$\text{APPROX-2}(L) = \text{REA}(L) - \sum_{i<j} \mathbb{P}[C_i \cup C_j]$$

This approximation under-estimates the top event probability. The reason for this lies in the fact that a status vector can fail several unions of pairs of minimal cut sets. This status vector will be then removed several times, potentially as many times as it has been added by REA or even more often. Those status vectors have to be added (included) again. We can do this by adding probabilities of unions of triplets of minimal cut sets and obtain the *third order approximation*, denoted APPROX-3, which is again a conservative estimate of the top event probability.

$$\text{APPROX-3}(L) = \text{APPROX-2}(L) + \sum_{i<j<k} \mathbb{P}[C_i \cup C_j \cup C_k]$$

We can continue this sequence of approximations including and excluding combinations of more and more cut sets:

$$\texttt{APPROX-(n+1)}(L) = \texttt{APPROX-n}(L) + (-1)^n \sum_{i_1 < \cdots < i_n} \mathbb{P}[C_{i_1} \cup \cdots \cup C_{i_n}]$$

These give alternating under- and overapproximations of the top event probability, converging to the true probability [1]:

$$\forall_{n \in \mathbb{N}} \texttt{APPROX-(2n)}(L) \leq \mathbb{P}[Top] \leq \texttt{APPROX-(2n+1)}(L)$$
$$\forall_{n \geq |L|} \texttt{APPROX-n}(L) = \mathbb{P}[Top]$$

The computational complexity of this process makes it impractical or even computationally infeasible. For an n-th approximation, we need to build $\frac{|L|!}{n! \cdot (|L|-n)!}$ unions of n-tuples of minimal cut sets, which asymptotically gives $|L|^n$ operations. In practice, software tools might calculate or approximate the second and possibly the third order approximation to evaluate the conservatism of the first order approximations (REA, or MCUB).

Binary Decision Diagrams. A minimal cut set list can be seen as a fault tree with a very special form. One can also encode it in a binary decision diagram (BDD) and use this data structure to quantify it. Depending on the variant of the BDD, we obtain the exact top event probability or an approximation. Some algorithms offer the possibility to determine the trade-off between the precision and the calculation complexity, both in terms of time and space. The method for building the BDD can make use of the special form of the fault tree to optimize the performance. Chapter 10 details quantification of fault trees by BDDs.

Appendix: Mathematical Proofs

Lemma 1 *Assume that basic events are stochastically independent. Let C_1 and C_2 be two cut sets of a fault tree F. Then, $\mathbb{P}[\neg C_1 \mid \neg C_2] \geq \mathbb{P}[\neg C_1]$.*

Proof

$$\begin{aligned}
\mathbb{P}[C_1 \mid C_2] &= \frac{\mathbb{P}[C_1 \cup C_2]}{\mathbb{P}[C_2]} \\
&= \frac{\mathbb{P}[(C_1 \setminus C_2) \cup (C_1 \cap C_2) \cup (C_2 \setminus C_1)]}{\mathbb{P}[(C_1 \cap C_2) \cup (C_2 \setminus C_1)]} \\
&= \frac{\mathbb{P}[C_1 \setminus C_2] \cdot \mathbb{P}[C_1 \cap C_2] \cdot \mathbb{P}[C_2 \setminus C_1]}{\mathbb{P}[C_1 \cap C_2] \cdot \mathbb{P}[C_2 \setminus C_1]} \quad \textit{(Independence of basic events)} \\
&= \mathbb{P}[C_1 \setminus C_2] \\
&\geq \mathbb{P}[C_1]
\end{aligned}$$

By the same argument, $\mathbb{P}[C_2 \mid C_1] \geq \mathbb{P}[C_2]$.

$$
\begin{aligned}
\mathbb{P}[\neg C_1 \mid \neg C_2] &= 1 - \mathbb{P}[C_1] \frac{\mathbb{P}[\neg C_2 \mid C_1]}{\mathbb{P}[\neg C_2]} && \text{(Bayes' rule)} \\
&= 1 - \mathbb{P}[C_1] \frac{1 - \mathbb{P}[C_2 \mid C_1]}{1 - \mathbb{P}[C_2]} && \\
&\geq 1 - \mathbb{P}[C_1] && \text{(from } \mathbb{P}[C_2 \mid C_1] \geq \mathbb{P}[C_2]\text{)} \\
&= \mathbb{P}[\neg C_1]
\end{aligned}
$$

□

Theorem 1 *Let $L = C_1, \ldots, C_n$ be the set of all minimal cut sets of a fault tree F, and let $\mathbb{P}[F]$ be the top probability of F. Then we have*

$$\mathbb{P}[F] \leq \text{MCUB}(L)$$

Proof

$$
\begin{aligned}
\mathbb{P}[F] &= \mathbb{P}[C_1 \vee C_2 \vee \ldots \vee C_n] \\
&= \mathbb{P}[\neg(\neg C_1 \wedge C_2 \wedge \ldots \wedge \neg C_n)] \\
&= 1 - \mathbb{P}[\neg C_1 \wedge \neg C_2 \wedge \ldots \wedge \neg C_n] \\
&= 1 - \mathbb{P}[\neg C_1] \cdot \mathbb{P}[\neg C_2 \mid \neg C_1] \cdot \ldots \cdot \mathbb{P}[\neg C_n \mid \neg C_1 \wedge \neg C_2 \wedge \ldots \wedge \neg C_{n-1}] \\
&\qquad\qquad \text{(Chain rule)} \\
&\leq 1 - \mathbb{P}[\neg C_1] \cdot \mathbb{P}[\neg C_2] \cdot \ldots \cdot \mathbb{P}[\neg C_n] && \text{(From Lemma 1)} \\
&= 1 - \prod_{C \in L} (1 - \mathbb{P}[C]) \\
&= \text{MCUB}(L)
\end{aligned}
$$

□

Theorem 2 *Let L be the set of all minimal cut sets of a fault tree F. Then we have*

$$\mathbb{P}[F] \leq \text{MCUB}(L) \leq \text{REA}(L)$$

Proof Let us call $L = \{C_1, C_2, \ldots, C_n\}$. We now prove

$$1 - \prod_{i=1}^{n}(1 - \mathbb{P}[C_i]) \leq \sum_{i=1}^{n} \mathbb{P}[C_i]$$

using induction on n:

- For $n = 1$, clearly $1 - (1 - \mathbb{P}[C_1]) = \mathbb{P}[C_1]$

- For $n > 1$:

$$1 - \prod_{i=1}^{n}(1 - \mathbb{P}[C_i])$$

$$= 1 - (1 - \mathbb{P}[C_n]) \prod_{i=1}^{n-1}(1 - \mathbb{P}[C_i])$$

$$= 1 - \left(\left(\prod_{i=1}^{n-1}(1 - \mathbb{P}[C_i]) \right) - \mathbb{P}[C_n] \prod_{i=1}^{n-1}(1 - \mathbb{P}[C_i]) \right)$$

$$= \left(1 - \prod_{i=1}^{n-1}(1 - \mathbb{P}[C_i]) \right) + \mathbb{P}[C_n] \prod_{i=1}^{n-1}(1 - \mathbb{P}[C_i])$$

$$\leq \left(\sum_{i=1}^{n-1}\mathbb{P}[C_i] \right) + \mathbb{P}[C_n] \prod_{i=1}^{n-1}(1 - \mathbb{P}[C_i]) \qquad \text{by induction hypothesis}$$

$$\left(\sum_{i=1}^{n-1}\mathbb{P}[C_i] \right) + \mathbb{P}[C_n] \prod_{i=1}^{n-1}1 \qquad \text{by } (1 - \mathbb{P}[C_i]) \in [0, 1]$$

$$\sum_{i=1}^{n}\mathbb{P}[C_i]$$

□

Combining both results above yields the following corollary.

Corollary 1 *Let L be the set of all minimal cut sets of a fault tree F. Then we have*

$$\mathbb{P}[F] \leq MCUB(L) \leq REA(L)$$

Reference

1. Ross SM (2007) Method of inclusion and exclusion, 9th edn. Academic Press, Chap. 9.4.1

Progress in Mathematical Physics

Computing Minimal Cut Sets

9

Cut set analysis requires a list of minimal cut sets of the fault tree to be available. Numerous algorithms exist to accomplish this task. The two main classes are based on *manipulation of lists of events*, and methods based on *binary decision diagrams (BDD)*. Other methods exploit, for example, SAT-solving and Monte Carlo simulation.

The list-manipulation methods include the classical MOCUS algorithm, a top-down approach that starts from top level event and builds the list downward. BDD-based methods rely on the compact BDD representation of the fault tree's structure function. These BDDs can be reused for other purposes, such as probabilistic analysis.

For very large fault trees, computing all minimal cut sets can be infeasible. In that case, approximation algorithms are used, which truncate low-probability cut sets.

Importantly, we note that the number of minimal cut sets in a fault tree can be exponential in the number of basic events, as illustrated in Fig. 9.1. Therefore, no algorithm performs fast in all cases.

Notation. In this section, we often use cut sets for events that are not the top event. A cut set for event E is a set of basic events that make E fail, i.e., a set $C \subseteq BE$ such that $\Phi_F(C, E) = 1$. A cut set C is a minimal cut set for E if removing any element from C makes C no longer a cut set.

By definition, a cut set is a set of basic events, denoted by curly brackets {..}. A set of cut sets is then a set of sets, or family of sets, e.g., {{Con, Eng}, {Bat, Eng}}. Cut set algorithms process these sets by representing them as lists, denoted by [..]. Thus, [Con, Eng] represents a cut set, and [[Con, Eng], [Bat, Eng]] represents a family of cut sets. For each event E, we write $M(E)$ for its family of minimal cut sets and $L(E)$ for the list representation of all cut sets. Given two lists L and L', we write $L{+}{+}L'$ for their concatenation, where L' is appended to L.

© The Author(s), under exclusive license to Springer Nature Switzerland AG 2026   97
M. Stoelinga et al., *Concise Guide to Fault Tree Analysis*, Computer Science Foundations and Applied Logic, https://doi.org/10.1007/978-3-031-78287-9_9

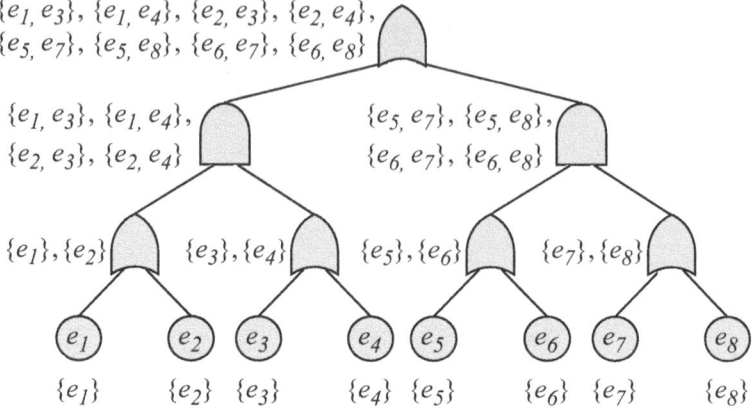

Fig. 9.1 Example illustrating the exponential growth of minimal cut sets

9.1 Set Manipulation Methods: MOCUS and MICSUP

Set manipulation methods use the fault tree structure to manipulate lists of events. We discuss two prominent methods: the top-down MOCUS and the bottom-up MICSUP algorithms.

9.1.1 Principles Behind MOCUS and MICSUP

Both MOCUS and MICSUP rely on the following principles, which obtain the minimal cut sets of an event from the minimal cuts sets of its children.

- Each basic event e has only one cut list, namely $\{e\}$.
- For events $E = \mathsf{OR}(E_1, \ldots, E_m)$ labeled with an OR-gate, we have that each minimal cut set C of any input E_i is also a cut set of E. However, C need not be minimal, since another child E_j could have a cut set that is smaller than C. Thus, the minimal cut sets of E are all minimal cut sets of its children, after removing the non-minimal ones.
- For events $E = \mathsf{AND}(E_1, \ldots, E_m)$ labeled with an AND-gate, we have that their minimal cut sets arise by picking one minimal cut set C_i for each input E_i, and taking their union $C_1 \cup C_2 \ldots \cup C_m$, where, again, we remove all non-minimal sets thus obtained.

We present the MOCUS and MICSUP algorithms for fault trees with AND- and OR-gates. VOTING-gates work similarly to the AND-gate, except that one first selects k children of E, ranging over all possible selections. Thus, VOTING-gates do give a blow up in the number of cut sets. Specialized solutions exist for this case [1].

9.1.2 The MOCUS Algorithm

MOCUS [2,3] stands for Method for Obtaining Cut Sets and is widely used in the field. MOCUS is a top-down implementation of the principles outlined above. That is, MOCUS starts at the top event. It then recursively explores the fault tree, expanding intermediate events according to the principles above, until the basic events are reached. The algorithm is illustrated in Example 16. Its code is given in Algorithm 1.

Example 16 (MOCUS algorithm to compute minimal cut sets)

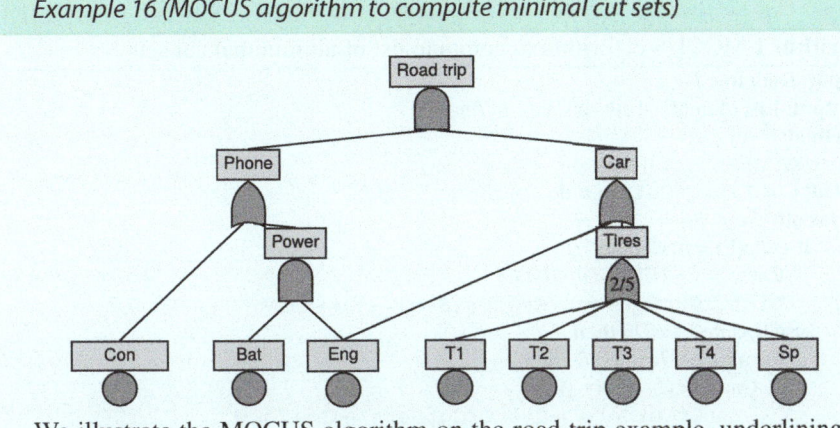

We illustrate the MOCUS algorithm on the road trip example, underlining the intermediate event that gets expanded.

- [[Road trip]], replace AND-gate by its children Phone, Car
- [[Phone, Car]], OR-gate: create copy for each of its children Con, Power
- [[Con, Car], [Power, Car]], replace AND-gate by its children
- [[Con, Car], [Bat, Eng, Car]], create copy for each of its children
- [[Con, Eng], [Con, Tires], [Bat, Eng, Car]]
- [[Con, Eng], [Con, Tires], [Bat, Eng, Eng], [Bat, Eng, Tires]]
- simplify to [[Con, Eng], [Con, Tires], [Bat, Eng], [Bat, Eng, Tires]]
- simplify to [[Con, Eng], [Con, Tires], [Bat, Eng]]
- [[Con, Eng], [Con, T1, T2],   [Con, T1, T3],   [Con, T1, T4], [Con, T1, Sp], [Con, T2, T3],   [Con, T2, T4], [Con, T2, Sp],   [Con, T3, T4], [Con, T3, Sp], [Con, T4, Sp], [Bat, Eng]]

Steps in MOCUS. Concretely, the MOCUS algorithm manipulates ListOfCutSets, a list of event lists. These events can be either intermediate or basic events. Initially, it contains one event list, namely [*Top*]. After termination, all lists in ListOfCutSets contain only basic events, namely, ListOfCutSets is exactly the list of all minimal cut sets of the fault tree.

The algorithm proceeds by replacing the intermediate events in ListOfCutSets by its inputs, as follows. For each intermediate event E that is contained in some list L in ListOfCutSets, the following steps are taken:

- If E is connected to an AND-gate with children $E_1, E_2, \ldots E_m$, then E is replaced by its children $E_1, E_2, \ldots E_m$, denoted by *Inputs(E)*.
- If E is connected to an OR-gate with children $E_1, E_2, \ldots E_m$, then n copies of L are created, where in copy i, event E is replaced by the event E_i. We remove the original list L from `ListOfCutSets`.
- Finally, Boolean laws are used to remove redundancies from the cut sets.

Algorithm 1 MOCUS algorithm to compute list of all minimal cut sets

---

Input: Fault tree F
Output: List of minimal cut sets, i.e., $L(Top)$
Method:
`ListOfCutSets` := [[*Top*]]
for all $C \in$ `ListOfCutSets` **do**
   for all $E \in C$ **do**
     if *Gate*(E) = AND **then**
       Cnew := C - [E] ++ *Inputs(E)*
       `ListOfCutSets` := `ListOfCutSets` -- C ++ Cnew
     else if *Gate*(E) = OR **then**
       for all $E' \in Inputs(E)$ **do**
         Cnew := C - [E] ++ [E']
         `ListOfCutSets` := `ListOfCutSets` -- C ++ Cnew
       end for
     end if
   end for
end for
return `ListOfCutSets`

9.1.3 The MICSUP Algorithm

The MICSUP (Minimal Cut Sets Upward) algorithm is a bottom-up variant of the MOCUS algorithm. The MICSUP method uses the same substitution and expansion principles as MOCUS, except that the operation begins at the bottom of the tree and proceeds upward. MICSUP has the advantage of providing the minimal cut sets not only for the top event, but also for all intermediate events. The algorithm is illustrated in Example 17.

Writing $L(E)$ for the set of all minimal cut sets of event E, the following relations hold, providing the steps in a recursive algorithm:

$$L(e) = [[e]] \qquad\qquad\qquad\qquad \text{if } e \text{ is a basic event}$$
$$L(E) = reduce(L(E_1) {+\!\!+} \ldots {+\!\!+} L(E_m)) \qquad\quad \text{if } E = OR(E_1, \ldots, E_m)$$
$$L(E) = reduce([C_1 {+\!\!+} \ldots {+\!\!+} C_m \mid C_i \in L(E_i)]) \text{ if } E = AND(E_1, \ldots, E_m)$$

Here, the function *reduce* takes a list of cut sets and removes all non-minimal cut sets. Reduction is a complex function, where heuristics are used to determine how and when to reduce. Strictly speaking reduction is not needed after each step.

Example 17 (MICSUP algorithm to compute minimal cut sets)

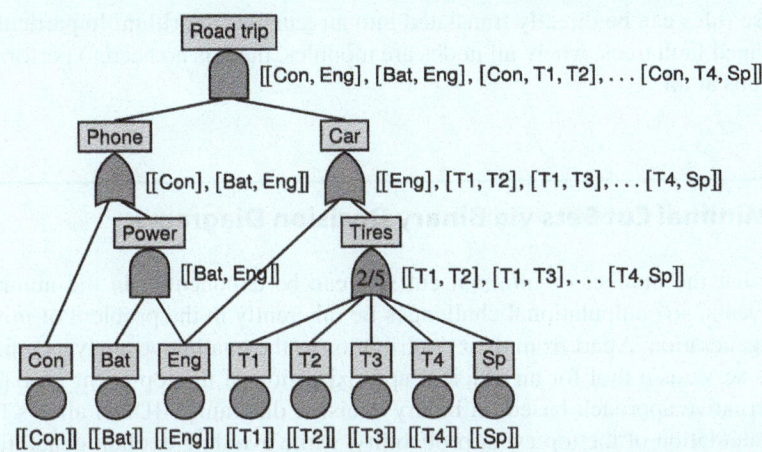

The figure above illustrates the MICSUP algorithm.

- We start at the bottom of the tree, setting $L(e) = [[e]]$.
- Since **Power** is equipped with an AND-gate, its cut sets arise by concatenating one minimal cut set from its child **Bat** and one from its child **Eng**. Both children have only one cut set, so we obtain $L(\text{Power}) = [[\text{Bat}]++[\text{Eng}]] = [[\text{Bat, Eng}]]$. No reduction is needed.
- $L(\text{Phone})$ is equipped with an OR-gate. Therefore, we have $L(\text{Phone}) = L(\text{Con})++L(\text{Power}) = [[\text{Con}]]++[[\text{Bat, Eng}]] = [[\text{Con}], [\text{Bat, Eng}]]$.
- For the 2/5 voting gate at **Tires**, we take any two combinations of tires. $[[\text{T1, T2}], [\text{T1, T3}], \ldots, [\text{T4, Sp}]]$.
- Since **Car** is equipped with an OR-gate $L(\text{Car}) = L(\text{Eng})++L(\text{Tires})$. Thus, we take all cut sets from both children: $L(\text{Car}) = [[\text{Eng}], [\text{T1, T2}], [\text{T1, T3}], \ldots, [\text{T4, Sp}]]$.
- For the AND-gate as the top, we combine each cut set of **Phone** with each cut set of **Car**. Here, we need to reduce: If we concatenate [Bat, Eng] with [Eng], we obtain [Bat, Eng, Eng], which should be reduced to [Bat, Eng].

MISCUP for Modules. Recall from Sect. 3.3 that a module is a node in a fault tree that represents an independent subtree. An event whose children are all modules does not require a reduction step. More precisely, if all children $E_1, \ldots, E_m$ of some event E are modules, their minimal cut sets are mutually disjoint. That is, for $i \neq j$, if $C \in L(E_i)$ and $C' \in L(E_j)$, then $C \cap C' = \emptyset$. Therefore, the following relations hold:

$$L(E) = L(E_1)++\ldots++L(E_m) \qquad \text{if } E = \mathsf{OR}(E_1, \ldots, E_m)$$
$$L(E) = [C_1++\ldots++C_m \mid C_i \in L(E_i)] \text{ if } E = \mathsf{AND}(E_1, \ldots, E_m)$$

These rules can be directly translated into a recursive algorithm. In particular, in tree-shaped fault trees, where all nodes are modules, there is no need to perform any reductions at all.

9.2 Minimal Cut Sets via Binary Decision Diagrams

Recall that the number of minimal cut sets can be exponential in the number of basic events, so computational challenges lie inherently in the problem of minimal cut set generation. Apart from their contribution to the qualitative analysis, minimal cut sets serve as a tool for an efficient approximation of the top event probability. An alternative approach based on binary decision diagrams (BDDs) allows for an exact calculation of the top event probability, while avoiding explicit generation of all minimal cut sets.

BDDs represent a compact way to encode Boolean functions [6,7]. They gained popularity because they can in some situations very effectively manage large-scale Boolean formulas. BDDs are applied in many domains, such as hardware and software verification and synthesis.

Since the structure function of a fault tree is a Boolean function, BDDs may also be exploited to encode the set of solutions, i.e., the set of cut sets of a fault tree in a symbolic manner [8,9]. One can obtain the top event probability and various other dependability metrics from this symbolic representation [9–11] without the need to list all minimal cut sets. But it is also possible to generate (minimal) cut sets from the resulting BDD to obtain qualitative insights.

Section 3.2 introduced the concept of BDD, as a compact representation of the structure function of a fault tree. Below, we show how to obtain the list of all cut sets as well as how to compute the list of all minimal cut sets.

9.2.1 Computing Cut Sets from a BDD

It is easy to read off the cut sets from a BDD. One starts at all 1-leaves of the tree, and traverses all paths upwards toward the root. Each path yields a different cut set.

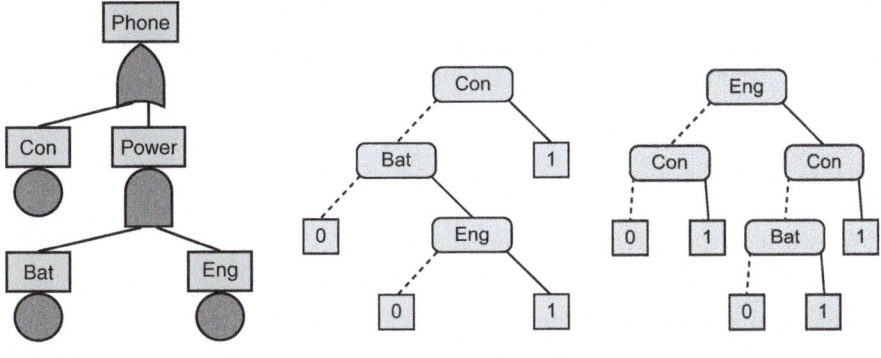

(a) Fault tree fragment   (b) Ordering Con < Bat < Eng   (c) Ordering Eng < Con < Bat

Fig. 9.2 BDDs with two different variable orderings

Example 18

Figure 9.2 shows the conversion of a fault tree into a BDD. By traversing the path leading to a 1 node, we obtain:

- The BDD in Fig. 9.2b yields as cut sets: {Bat, Eng} and {Con}.
- The BDD in Fig. 9.2c yields as cut sets: {Con}, {Bat, Eng} and {Con, Eng}.

All these are cut sets, but the last one is not minimal. The section below explains how to obtain a BDD representing only the minimal cut sets. Moreover, not all cut sets are paths in the BDD; for example, {Con, Bat, Eng} is a cut set, but does not correspond to a BDD path.

9.2.2   Computing All Minimal Cut Sets using a BDD

As shown above, the BDD obtained via the Shannon expansion produces a list of cut sets that cover the whole failure behavior of the fault tree; however, the cut sets in this list do not need to be minimal. This is illustrated in Fig. 9.3

To obtain minimal cut sets, one needs to adapt the Shannon expansion [9] using the formula

$$f_{\min}(x_1, x_2, \cdots, x_n) = (\neg x_1 \wedge f_{\min}(0, x_2, \cdots, x_n))$$
$$\vee (x_1 \wedge f_{\min}(1, x_2, \cdots, x_n) \wedge \neg f(0, x_2, \cdots, x_n))$$

To understand the rationale behind this formula, we observe that the status vector $x_1, x_2, \ldots, x_n$ represents a minimal cut set in the following cases:

- If $x_1 = 0$, then $x_2, \ldots, x_n$ must represent a minimal cut set.
- If $x_1 = 1$, then $x_2, \ldots, x_n$ must represent a minimal cut set, and moreover, $0, x_2, \ldots, x_n$ must not be a cut set. Indeed, if $0, x_2, \ldots, x_n$ is a cut set—either minimal or not—then $1, x_2, \ldots, x_n$ is not a minimal cut set.

This is exactly what is expressed in the formula above.

One can compute the BDD for $f_{\min}$ via its Shannon expansion, or from the formula above by using the difference operator on BDDs. Alternatively, one can construct a BDD encoding the prime implicants. We refer the reader to [12] for more details.

Heuristics for variable ordering. While the conversion to a BDD has exponential worst-case complexity, it has linear complexity in the best case, and show excellent performance in practice (Fig. 9.4). This is strongly influenced by the fact that BDDs very compactly represent Boolean functions with a high degree of symmetry [13], and fault trees exhibit this symmetry as the gates are symmetric in their inputs.

Remenyte and Andrews [14, 15] have compared several different methods for constructing BDDs from fault trees and conclude that a hybrid of the if-then-else method [9] and the advanced component-connection method by Way and Hsia[16] is a good trade-off between processing time and size of the resulting BDD.

9.2.3 Improvements to BDD

Tang and Dugan[17] propose the use of zero-suppressed BDDs to compute minimal cut sets. This approach is more efficient than those based on classic BDDs in both time and memory use.

If subtrees of a fault tree are shared, then the approach by Codetta-Raiteri[18] called "Parametric Fault Trees" can be used. This method performs qualitative and quantitative analysis on such trees without repeating the analysis for each repetition of a subtree.

Fig. 9.3 BDD obtained via Shannon expansion with non-minimal cut set {Con, Eng}

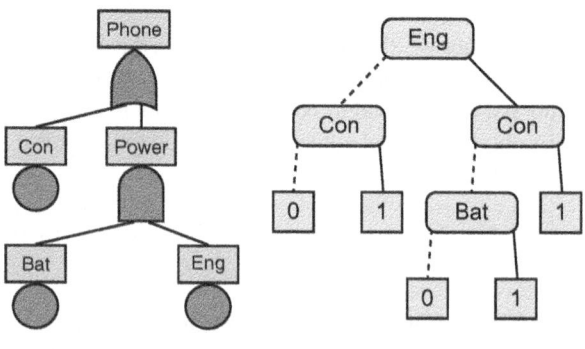

(a) Fault tree fragment   (b) Ordering Eng < Con < Bat

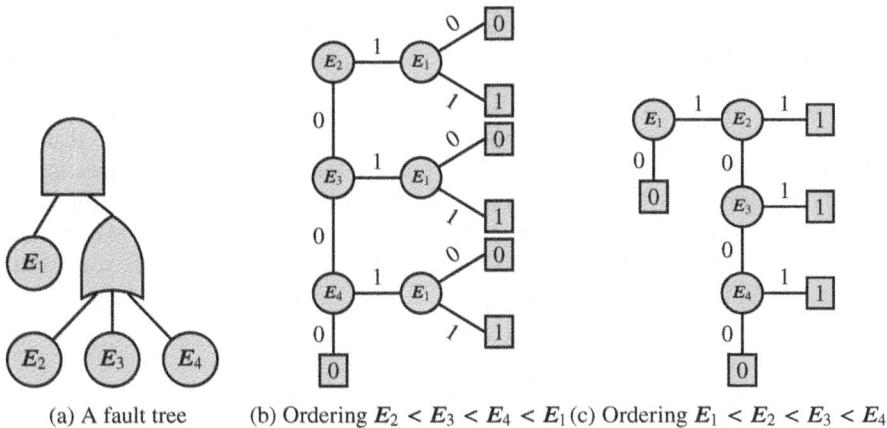

(a) A fault tree   (b) Ordering $E_2 < E_3 < E_4 < E_1$ (c) Ordering $E_1 < E_2 < E_3 < E_4$

Fig. 9.4 Example of how variable ordering affects BDD size: **b** has 13 vertices, while **c** has 9. Other orderings are possible, but are not obvious

9.3   Key Techniques for Efficiency Improvement

The descriptions of the minimal cut set algorithms presented above lay out their structure and main steps, but they lack the critical techniques needed for efficient implementation, leaving a significant opportunity for optimization and performance enhancement. Incorporating these techniques can drastically improve computational efficiency and make the algorithms far more practical for real-world applications.

Fault Tree Preprocessing. A fault tree created by an analyst will typically have a structure that supports easy understanding of the model by other analysts or reviewers. This structure might be less suited for generating minimal cut sets. The fault tree can be typically simplified and restructured to maximize reuse of its parts. A good preprocessing will also influence following steps.

Modularization. Even if the fault tree to be analyzed has the form of a directed acyclic graph, some of its parts might be trees. For the purpose of minimal cut set generation, these trees can be replaced by new basic events which represent the whole tree, so called *modules* [19]. Modules were described in Sect. 3.3 and, as pointed out in Sect. 9.1.3, modularization can significantly simplify the MICSUP algorithm. However, also the MOCUS and BDD method can be combined with modularization. This process reduces the size of the fault tree. The minimal cut set list obtained from a modularized fault tree will contain modules. They must be replaced by basic event combinations from the original trees corresponding to the modules. One can use the MOCUS algorithm for this process as well.

Cut Set Generation. In their plain form, the algorithms for generating cut sets can duplicate work if a shared event occurs in multiple cut sets. To avoid this duplication of work, the efficiency of these algorithms crucially relies on carefully chosen data structures. For MOCUS and MICSUP, encoding the search state information requires either custom developed or generic ones such as binary decision trees or zero-suppressed BDDs [3]. BDD algorithms require good hashing and reuse, together with smart heuristics for variable ordering [20, 21].

Minimization. An important step in the MOCUS algorithm discards cut sets that are not minimal. The earlier in the generation process the algorithm recognizes that a partially developed cut set will not lead to any minimal one, the higher the efficiency will be. For coherent fault trees, a simple test can determine if a cut set is minimal. We remove one basic event at a time and check if the cut set still fails the top event. If it is the case, this cut set is not minimal.

9.4 Truncation

Since the number of minimal cut sets may grow exponentially with the number of basic events, cut set analysis for large fault trees is complex, and often even infeasible, both in terms of computation and their interpretation. A key technique to handle minimal cut sets for large fault tree is via *truncation*, also called *cutoff*. This technique ignores minimal cut sets whose influence on the top probability is assessed to be small, that is, whose probability is very low.

The rationale behind truncation is that systems assessed by fault tree analysis often comprise dependable components with low failure rates, featuring basic event failure probabilities between $1 \cdot 10^{-2}$ and $1 \cdot 10^{-6}$.

Thus, the cut set probability drops quickly with the number of events in the set, so that higher-order cut sets will only marginally impact the overall value. Moreover, beyond a certain precision, the model uncertainties will outweigh the additional decimal points obtained by considering high-order cut sets.

> ### Example 19 (Truncation of minimal cut sets in the road trip example)
>
> Table 9.1 recalls the cut set probabilities from Example 11, where we summarize the 10 cut sets containing Con and two tires in one row. The third column shows the effect of restricting the minimal cut sets to order two or less. Restricting the cut set probability to (for example) at least 5.0E-3 yields the same results.
>
> Recall from Table 7.2 that the total cut set probability equals 4.75E-2. The two dominant cut sets {Con, Eng} and {Bat, Eng} have a combined probability of 1.25E-2 + 1.0E-2 = 2.25E-2. The probability of the remaining cut sets sums to 4.75E-2 − 2.25E-2 = 2.5E-2. In this case, generating also the cut sets of order three is definitely worth the effort. The situation might be very different if the failure probability of a tire drops to 0.01. Then the truncation removes less than two percent of the top value.

Restricting the search for minimal cut sets to those that significantly impact the top event value accelerates the analysis process. The most common constraints involve

Table 9.1 Influence of dominant cut sets

MCS	All MCSs	Up to order 2
Con, Eng	$0.25 \cdot 0.05 = 1.25\text{E--}2$	1.25E–2
Bat, Eng	$0.2 \cdot 0.05 = 1.0\text{E--}2$	1.0E–2
Con, Ti, Tj	$10 \cdot 0.25 \cdot 0.1 \cdot 0.1 = 2.5\text{E-}3$	
Total	$= 4.75\text{E-}2$	$= 2.25\ \text{E-}2$

setting a probability threshold for cut sets (a "cutoff") or limiting the order of the cut sets. However, because this approach results in an underestimation of the total number of cut sets, which in turn may be conservatively quantified by a Rare Event Approximation or a Min Cut Upper Bound, this technique cannot provide any guaranteed bounds on the top event probability.

However, in practical applications, selecting a cutoff based on experience or established heuristics often yields highly accurate results, especially considering uncertainties in the failure data. One way to validate the chosen cutoff is to perform the analysis with a cutoff value ten times lower and compare the resulting top event probabilities, as well as estimate any cutoff-induced error. Future research could explore methods to better estimate the impact of truncation on importance measures.

For example, if the expected top value is around 10^{-4}, then a typical cutoff value might be 10^{-10} for a quick analysis that provides informative, though not exceptionally precise, results. For highly accurate results while maintaining practical analysis times, a cutoff in the range of 10^{-14} to 10^{-15} is recommended. This approach offers a good balance between precision and efficiency compared to conducting an analysis without any cutoff.

9.5   Other Methods for Computing Cut Sets

9.5.1   Computation via SAT Solving

Several approaches exploit progress on solving the Boolean satisfiability problem (SAT) [22]. Given a Boolean formula, solving SAT means determining whether a satisfying assignment (or model) for that formula exists. Solving SAT is proven to be NP-complete [23]; however, SAT solvers are heavily used in practice, also to replace BDD-based procedures [24, 25].

SATMCS [26] is a method for computing MCSs based on SAT, which works as follows. First, the structure function of a fault tree is encoded in its conjunctive normal form (CNF) $\mathcal{F}$. The CNF can be generated in linear time, as it is not strictly equivalent to the original formula due to the introduction of new variables. Rather, it is equisatisfiable, meaning that for any given assignment of truth values, the new formula will be satisfied if and only if the original formula is satisfied.

SATMCS maintains sets of clauses $\mathcal{H} = \mathcal{F} \cup \mathcal{B} \cup \mathcal{C}$ and $\mathcal{M}$. As mentioned, $\mathcal{F}$ encodes the fault tree in CNF; then, using block clauses, $\mathcal{B}$ blocks already computed minimal cut sets recorded by $\mathcal{M}$; and $\mathcal{C}$ contains conflict clauses (obtained from the conflict-driven clause learning (CDCL) framework [27]). When iterating, SATMCS uses $\mathcal{C}$ to continuously reduce the search space in which there is no model (i.e., there is no cut set) for the fault tree. At the same time, SATMCS uses $\mathcal{B}$ to block the models that has been found. Eventually, $\mathcal{B}$ prunes all the search space covered by all minimal cut sets, meaning that all the minimal cut sets for the fault tree have been found. $\mathcal{H}$ is thus unsatisfiable, and the algorithm terminates.

Quantitative SAT can be used to compute Maximum Probability Minimal Cut Sets (MPMCSs) [28]. Their method uses MaxSAT [29], which finds a truth assignment that maximizes the weight of satisfied clauses. To represent the non-occurrence of the top level event, they negate the structure function: basic events are negated, and logic gates are flipped (AND becomes OR, and vice versa). This process maximizes falsified variables and minimizes satisfied ones, forming a minimal cut set in the fault tree [28]. The formula is then encoded in CNF, and the probabilities of BEs are transformed using a negative-log function. This transformation allows MaxSAT to maximize the product of weighted decision variables. Soft clauses are created for each decision variable, with penalties for falsification. The MaxSAT solver minimizes the total weight of falsified variables, resulting in a minimum vertex cut in logarithmic space, corresponding to the highest probabilities and identifying the MPMCS–MCS with maximum joint probability [28].

9.5.2 Algorithms for Special Cases

For fault trees with voting gates with many inputs, a combinatorial explosion can occur, since a k/N voting gate means each combination of k failed components results in a separate cut set. The concept of a Minimal Cut Vote can be utilized to represent an arbitrary combination of k elements [1] .

For relatively large trees with few cut sets, the algorithm by Carrasco and Suñé [30] may be useful. Its space complexity is based on the minimal cut sets, rather than the complexity of the tree like for BDDs. However, according to the article this method does seem to be slower than the BDD approach.

9.6 Computing Path Sets

Any algorithm to compute minimal cut sets can also be used to compute minimal path sets. To do so, the fault tree is replaced by its dual, called the success tree. Here AND-gates are replaced by OR-gates, OR-gates by AND-gates, k/N-voting gates by $(N - k + 1)/N$-voting gates, and basic events by their complement (i.e., "component failure" by "no component failure"). The minimal cut sets of the success tree are exactly the minimal path sets of the original fault tree [31].

References

1. Xiang J, Yanoo K, Maeno Y, Tadano K, Machida F, Kobayashi A, Osaki T (2011) Efficient analysis of fault trees with voting gates. In: Proceedings of the 22nd IEEE international symposium on software reliability engineering (ISSRE), IEEE, pp 230–239. https://doi.org/10.1109/ISSRE.2011.23

2. Fussell JB, Vesely WE (1972) New methodology for obtaining cut sets for fault trees. Trans Am Nucl Soc

3. Rauzy A (2003) Toward an efficient implementation of the MOCUS algorithm. IEEE Trans Reliab 52(2):175–180. https://doi.org/10.1109/TR.2003.813160

4. Walker MD (2009) Pandora: a logic for the qualitative analysis of temporal fault trees. PhD thesis, University of Hull

5. Vesely WE, Goldberg FF, Roberts NH, Haasl DF (1981) Fault tree handbook. Nuclear Regulatory Commission, Office of Nuclear Regulatory Research, U.S

6. Akers SB (1978) Binary decision diagrams. IEEE Trans Comput C-27(6):509–516. https://doi.org/10.1109/TC.1978.1675141

7. Bryant RE (1986) Graph-based algorithms for Boolean function manipulation. IEEE Trans Comput C-35(8):677–691. https://doi.org/10.1109/TC.1986.1676819

8. Coudert O, Madre JC (1993) Fault tree analysis: 10^{20} Prime implicants and beyond. In: Proceedings of the reliability and maintainability symposium (RAMS). IEEE, pp 240–245. https://doi.org/10.1109/RAMS.1993.296849

9. Rauzy AB (1993) New algorithms for fault tree analysis. Reliab Eng Syst Saf 40(3):203–211. https://doi.org/10.1016/0951-8320(93)90060-C

10. Sinnamon RM, Andrews JD (1996) Fault tree analysis and binary decision diagrams. In: Proceedings of the reliability and maintainability symposium (RAMS). IEEE, pp 215–222. https://doi.org/10.1109/RAMS.1996.500665

11. Basgöze D, Volk M, Katoen J, Khan S, Stoelinga M (2022) BDDs strike back—efficient analysis of static and dynamic fault trees. In: Deshmukh JV, Havelund K, Perez I (eds) Proceedings of the 14th international symposium on NASA formal methods (NFM). Springer, Lecture Notes on Computer Science, vol 13260, pp 713–732

12. Rauzy A, Dutuit Y (1997) Exact and truncated computations of prime implicants of coherent and non-coherent fault trees within Aralia. Reliab Eng Syst Saf 58(2):127–144. https://doi.org/10.1016/S0951-8320(97)00034-3

13. Ross DE, Butler KM, Mercer MR (1991) Exact ordered binary decision diagram size when representing classes of symmetric functions. J Electron Test 2(3):243–259. https://doi.org/10.1007/BF00135441

14. Remenyte R, Andrews JD (2006) A simple component connection approach for fault tree conversion to binary decision diagram. In: Proceedings of the 1st international conference on availability, reliability and security (ARES), pp 449–456. https://doi.org/10.1109/ARES.2006.17

15. Remenyte-Prescott R, Andrews J (2008) An enhanced component connection method for conversion of fault trees to binary decision diagrams. Reliab Eng Syst Saf 93(10):1543–1550. https://doi.org/10.1016/j.ress.2007.09.001

16. Way YS, Hsia DY (2000) A simple component-connection method for building binary decision diagrams encoding a fault tree. Reliab Eng Syst Saf 70(1):59–70. https://doi.org/10.1016/S0951-8320(00)00048-X

17. Tang Z, Dugan JB (2004) Minimal cut set/sequence generation for dynamic fault trees. In: Proceedings of the reliability and maintainability symposium (RAMS). IEEE, pp 207–213. https://doi.org/10.1109/RAMS.2004.1285449

18. Codetta-Raiteri D (2006) BDD based analysis of parametric fault trees. In: Proceedings of the reliability and maintainability symposium (RAMS). IEEE, pp 442–449. https://doi.org/10.1109/RAMS.2006.1677414

19. Dutuit Y, Rauzy AB (1996) A linear-time algorithm to find modules of fault trees. IEEE Trans Reliab 45:422–425. https://doi.org/10.1109/24.537011
20. Bouissou M, Bruyere F, Rauzy A (1997) BDD based fault-tree processing: a comparison of variable ordering heuristics. In: European safety and reliability association conference. ESREL
21. Rauzy A (2008) Some disturbing facts about depth-first left-most variable ordering heuristics for binary decision diagrams. Proc Inst Mechan Eng Part O: J Risk Reliab 222(4):573–582
22. Biere A, Heule M, van Maaren H (2009) Handbook of satisfiability. IOS Press, vol 185
23. Cook SA (2023) The complexity of theorem-proving procedures. In: Logic, automata, and computational complexity: the works of Stephen A. Cook, Association for Computing Machinery, pp 143–152
24. Biere A, Cimatti A, Clarke E, Zhu Y (1999) Symbolic model checking without BDDs. In: Cleaveland W (ed) Proceedings of the international conference on tools and algorithms for the construction and analysis of systems (TACAS), Springer, Lecture Notes in Computer Science, vol 1579, pp 193–207. 10.1007/3-540-49059-0_14
25. Cimatti A, Pistore M, Roveri M, Sebastiani R (2002) Improving the encoding of LTL model checking into SAT. In: International workshop on verification, model checking, and abstract interpretation. Springer, pp 196–207
26. Luo W, Wei O, Wan H (2020) SATMCS: an efficient sat-based algorithm and its improvements for computing minimal cut sets. IEEE Trans Reliab 70(2):575–589
27. Eén N, Sörensson N (2003) An extensible SAT-solver. In: Giunchiglia E, Tacchella A (eds) International conference on theory and applications of satisfiability testing. Springer, Lecture Notes in Computer Science, pp 502–518. https://doi.org/10.1007/978-3-540-24605-3_37
28. Barrère M, Hankin C (2020) Fault tree analysis: identifying maximum probability minimal cut sets with MaxSAT. In: 2020 50th annual IEEE-IFIP international conference on dependable systems and networks-supplemental volume (DSN-S). IEEE, pp 53–54
29. Davies J, Bacchus F (2011) Solving MAXSAT by solving a sequence of simpler SAT instances. In: Lee J (ed) International conference on principles and practice of constraint programming (CP). Springer, Lecture Notes in Computer Science, vol 6876, pp 225–239. https://doi.org/10.1007/978-3-642-23786-7_19
30. Carrasco JA, né VS, (1999) An algorithm to find minimal cuts of coherent fault-trees with event-classes using a decision tree. IEEE Trans Reliab 48:31–41. https://doi.org/10.1109/24.765925
31. Barlow RE, Proschan F (1975) Statistical theory of reliability and life testing. Holt, Rinehart, & Winston

Part IV
Quantitative Fault Tree Analysis

Failure Probabilities via the Point-Probabilistic Model

10

Probabilistic fault tree analysis studies the failure behavior of fault trees in terms of probabilities. It starts with equipping all basic events with probabilistic data. Depending on the metric of interest, these events are decorated with failure probabilities, frequencies, rates, repair times, or inspection intervals.

In this chapter, we study an elementary model where each basic event is assigned a single failure probability, just as we did for cut sets. We call this the *point-probabilistic model*, since these probabilities represent the likelihood for the event to happen at one point in time. This point can be after the coming year, per hour, per operational cycle or when the demand for the required function arises.

From the basic event probabilities, we compute the failure probabilities of all other events, and especially the failure probability of the top event, called the top probability. The computation method strongly depends on the shape of the tree:

- Tree-shaped fault trees can be analyzed via a bottom-up method that propagates the failure probabilities along the tree.
- For DAG-shaped fault trees, we present a BDD-based method to compute these probabilities.

Whereas the cut set based methods from Chap. 8 allow us to approximate the top event probability, this chapter provides methods to calculate the top probability in a mathematically exact way. As usual, the quality of the results strongly depends on the quality of the input parameters. Therefore, uncertainties must be evaluated, for example, via Monte Carlo simulation or sensitivity analysis.

The next chapters (Chaps. 11 and 12) study more complex probabilistic models, where failure probabilities vary over time. These models equip the basic events with different probabilistic information, such as failure rates and/or repair rates.

Independence. It is crucial to realize that this probabilistic failure analysis assumes that all failures of basic events are probabilistically *independent*, that is, the failure of one basic event does not influence the failures of other basic events. If the basic events are not independent, then the analysis usually yields too optimistic outcomes. To model cases where failures are not independent, we refer the reader to Chap. 14 for the treatment of common causes.

Notation. In this chapter, we assume a fault tree F whose basic events are $e_1, e_2, \ldots,$ e_n. We denote its structure function Φ_F. Further, we equip each basic event with a single probability $\mathbb{P}[e]$, which is sometimes denoted Q_e in the literature.

10.1 Point Probabilities: Reliability Models for Basic Events

Before performing any computations, it is crucial to understand the source and nature of failure probabilities. Basic events represent atomic failures or failure modes of components, operators, infrastructure, or other factors that could lead to an accident or system-level failure. In this chapter, each basic event is linked to a single failure probability. These probabilities can be interpreted and obtained in various ways, depending on the context, such as whether a component is repairable. In all cases, a mathematical *reliability model* is used to capture the associated failure behavior.

These models help us determine the probability that a component will fail to perform its intended function. Failure occurs when the component is unavailable at the moment it is needed. We assume that the demand for this component arises randomly, following a uniform distribution.

One of the simplest ways to estimate failure probabilities is by determining the *failure rate*—the frequency at which a specific type of failure occurs within a given time interval. This applies to various types of components. For components in active operation, failures may occur, but a detection system alerts a repair crew to restore them with a certain *repair rate*. Other components are kept off as backups or for emergencies. When needed, they must first be started, and if an internal failure occurred while they were inactive, they will fail to start. These failures often go unnoticed until the component is tested, so periodic tests are conducted at specified *test intervals*. If a failure is detected during a test, the component is repaired immediately. Another scenario involves a component that starts successfully and operates for a designated period—known as *mission time*—during which a failure can occur at any moment.

Repairable Reliability Model. The probability of a basic event with a repairable reliability model reflects the unavailability of the component at a given random time point.

The component may either be operational when needed, in which case it succeeds, or it may have recently failed and is undergoing repair, resulting in its failure to perform its function. The failure probability is calculated as follows:

- Failure rate: λ.
- Repair rate: μ.
- Failure probability: $\mathbb{P}[e] = Q_e = \frac{\lambda}{\lambda + \mu}$.

Periodically Tested Reliability Model. When the demand comes, we try to start the component. It either starts successfully or it fails to start in which case the component fails to provide its function. The component is tested at periodic times, and immediately repaired if the component is non-functional. The failure probability can be calculated as follows:

- Failure rate: λ.
- Test interval: TI.
- Failure probability: $\mathbb{P}[e] = Q_e = 1 - \frac{1 - e^{-\lambda \cdot TI}}{\lambda \cdot TI}$.

Mission Time Reliability Model. When the demand comes, we assume that this component has just successfully started. It either operates successfully for the specified mission time or it fails in operation before the end of the mission time in which case this component fails to provide its function. The failure probability can be calculated as follows:

- Failure rate: λ.
- Mission time: T.
- Failure probability: $\mathbb{P}[e] = Q_e = 1 - e^{-\lambda \cdot T}$.

Probability Reliability Model. This model does not depend on the actual failure mechanism, as long as the mean unavailability (the failure probability) can be expressed by a single probability parameter.

- Probability: p.
- Failure probability: $\mathbb{P}[e] = Q_e = p$.

Each of these models produces a single probability value representing the probability that the component fails to perform its function when required. Different component reliability models can be used within one fault tree.

10.2 The Point-Probabilistic Model

Basic Event Probabilities. The reliability models yield for each basic event e a single probability $\mathbb{P}[e]$. We refer to this approach as the *point-probabilistic model*. The aim is to compute the failure probabilities for intermediate events, denoted by E, and in particular the *top probability* of the fault tree F, i.e., failure probabilities $\mathbb{P}[E]$ and $\mathbb{P}[F]$.

Instead of $\mathbb{P}[e]$, we also write p_e, or p_i if $e = e_i$. In this way, we obtain a vector of probabilities $\mathbf{p} = (p_1, \ldots, p_n)$ that describes the collective probabilistic behavior of all basic events in the fault tree. We assume that all failures of basic events $e_1, \ldots, e_n$ are mutually independent.

Probabilities of Status Vectors. Using the probabilistic vector $\mathbf{p}$, one can compute the probabilities of all other events via the structure function. First, we assign a probability to each status vector. Given a probabilistic vector $\mathbf{p} = (p_1, \ldots, p_n)$, the probability for a status vector $\mathbf{b} = (b_1, \ldots, b_n)$ is given by

$$\mathbb{P}[\mathbf{b}] = \mathbb{P}[b_1] \cdots \mathbb{P}[b_n] = \prod_{1 \le i \le n} b_i \cdot \mathbb{P}[e_i] + (1 - b_i) \cdot (1 - \mathbb{P}[e_i])$$

For example, given the probability vector $(p_1, p_2, p_3, p_4, p_5)$, the status vector 00101 is assigned probability $(1 - p_1)(1 - p_2)p_3(1 - p_4)p_5$. We often abbreviate $\overline{p} = 1 - p$, writing $\overline{p_1}\,\overline{p_2}\,p_3\,\overline{p_4}\,p_5$. Applying this notation to Boolean variables boils down to negation, i.e., $\overline{b} = \neg b$.

Probabilities of Intermediate and Top Events. The probability of the top and intermediate events E is obtained by summing over all status vector probabilities. That is, we add the probabilities of all status vectors $\mathbf{b}$ that make event E fail, i.e., with $\Phi_F(\mathbf{b}, E) = 1$. We define a probabilistic version of the status function, which takes as input the basic event probabilities and outputs the probability for E to fail.

$$\Phi_F^{\mathrm{p}}(p_1, \ldots, p_n, E) = \sum_{\mathbf{b} \in \{0,1\}^n} \mathbb{P}[\mathbf{b}] \cdot \Phi_F(\mathbf{b}, E)$$

Example 20 (Probabilistic fault tree)

As also shown in Example 5, the top probability is calculated as shown in the table, by summing the probabilities of all status vectors that result in failure of the top event.

Since this fault tree is tree-shaped, we can also calculate the probability as $p_{\mathrm{Con}} + p_{\mathrm{Bat}} \cdot p_{\mathrm{Eng}} - p_{\mathrm{Con}}\, p_{\mathrm{Bat}} \cdot p_{\mathrm{Eng}} = 0.2575$.

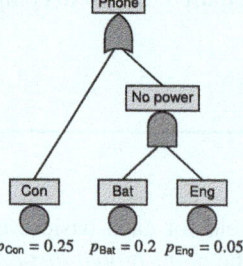

b	$\Phi_F(\mathbf{b})$	$\mathbb{P}[\mathbf{b}]$
000	0	
001	0	
010	0	
011	1	$\overline{p}_{\mathrm{Con}}\,p_{\mathrm{Bat}}\,p_{\mathrm{Eng}} = 0.75 \cdot 0.2 \cdot 0.05 = 0.0075$
100	1	$p_{\mathrm{Con}}\,\overline{p}_{\mathrm{Bat}}\,\overline{p}_{\mathrm{Eng}} = 0.25 \cdot 0.8 \cdot 0.95 = 0.19$
101	1	$p_{\mathrm{Con}}\,\overline{p}_{\mathrm{Bat}}\,p_{\mathrm{Eng}} = 0.25 \cdot 0.8 \cdot 0.05 = 0.01$
110	1	$p_{\mathrm{Con}}\,p_{\mathrm{Bat}}\,\overline{p}_{\mathrm{Eng}} = 0.25 \cdot 0.2 \cdot 0.95 = 0.0475$
111	1	$p_{\mathrm{Con}}\,p_{\mathrm{Bat}}\,p_{\mathrm{Eng}} = 0.25 \cdot 0.2 \cdot 0.05 = 0.0025$
		TOTAL = 0.2575

Phone — No power — Con — Bat — Eng

$p_{\mathrm{Con}} = 0.25$ $p_{\mathrm{Bat}} = 0.2$ $p_{\mathrm{Eng}} = 0.05$

Correctness. It can be shown that Φ_F^p correctly calculates the failure probabilities, i.e., that

$$\Phi_F^p(\mathbf{p}, E) = \mathbb{P}[\Phi_F(\mathbf{b}, E) = 1]$$

Moreover, Φ_F respects the gate logic, i.e., for each event E with children $E_1, \ldots, E_n$, it can be shown that

$$\Phi_F^p(\mathbf{p}, E) =$$
$$\begin{cases} \mathbb{P}[E_1 \text{ fails} \wedge \ldots \wedge E_n \text{ fails}] & \text{If } E \text{ is an AND-gate} \\ \mathbb{P}[E_1 \text{ fails} \vee \ldots \vee E_n \text{ fails}] & \text{If } E \text{ is an OR-gate} \\ \mathbb{P}[\text{There are } i_1, i_2, \ldots, i_k \text{ with: } E_{i_1}, \ldots, E_{i_k} \text{ fails}] & \text{If } E \text{ is a VOT}(k/N)\text{-gate} \end{cases}$$
$$(10.1)$$

Efficiency. As can be seen in Example 10.2, this method is not feasible for performing practical computations, since a fault tree with n basic events has 2^n status vectors. Hence, practical methods avoid enumerating all status vectors. Rather, bottom-up methods or BDDs are used, see Sects. 10.3 and 10.4.

10.3 Calculating Failure Probabilities in Tree-Shaped Fault Trees

In tree-shaped fault trees, the failure probabilities can be obtained via bottom-up computation: One starts with the probabilities at the basic events and propagates these upward in the tree, using appropriate rules to compute the gate probabilities. These rules are given in Table 10.1 and exemplified below.

Table 10.1 Rules for computing gate probabilities in tree-shaped fault trees

Symbol	Name	Probability
p_O $p_1 \cdots p_N$	AND-gate	$p_O = p_1 \cdot p_2 \cdots p_N$
p_O $p_1 \cdots p_N$	OR-gate	For 2 inputs: $p_O = p_1 + p_2 - p_1 p_2$ Multiple inputs: $p_O = 1 - (1 - p_1)(1 - p_2) \cdots (1 - p_N)$
p_O k/N $p_1 \cdots p_N$	VOT-gate	$p_O = p_1 p_{\text{VOT}}^{k-1, N-1}(p_2, \ldots p_N) + (1 - p_1) p_{\text{VOT}}^{k, N-1}(p_2, \ldots p_N)$ Here, For the special case that $p_1 = p_2 = \ldots = p_N = p$, we have: $p_O = \sum_{i=k}^{N} \binom{N}{i} p^i (1-p)^{N-i}$

Note that for tree-shaped fault trees it is not useful to express the voting gate as a combination of AND- and OR-gates, since that would destroy the tree structure of the fault tree. Therefore, we present a dedicated rule for the voting gate, which can be understood as a probabilistic version of a Shannon expansion. We elaborate these rules in Sect. 10.3.1, after illustrating them on two examples.

Example 21 (Bottom-up method for computing failure probabilities)

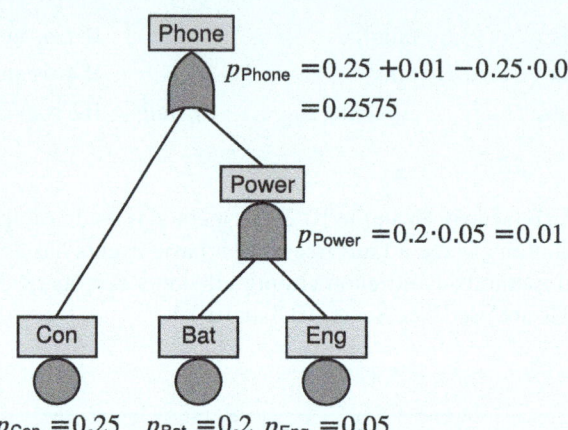

$p_{Phone} = 0.25 + 0.01 - 0.25 \cdot 0.01$
$= 0.2575$

$p_{Power} = 0.2 \cdot 0.05 = 0.01$

$p_{Con} = 0.25 \quad p_{Bat} = 0.2 \quad p_{Eng} = 0.05$

The bottom-up algorithm starts at the basic events and works its way up in the tree, using the rules in Table 10.1.

- **Power** carries an AND-gate, so we multiply its input probabilities: $p_{Power} = p_{Bat} \cdot p_{Eng} = 0.2 \cdot 0.05 = 0.01$.
- **Phone** is equipped with an OR-gate, so we obtain:
  $p_{Phone} = p_{Con} + p_{Power} - p_{Con} \cdot p_{Power} = 0.25 + 0.01 - 0.25 \cdot 0.01 = 0.2575$.

The top level event probability of the complete road trip example can not be computed via the bottom-up algorithm, since it is not a proper tree. Its calculation is explained in Sect. 10.4.

Example 22 (Running example: computing failure probabilities)

Finally, Car is equipped with an OR-gate again, and Tires is equipped with a VOT-gate. So we get $p_{\text{Tires}} = \sum_{i=2}^{5} \binom{5}{i} 0.1^i (1 - 0.1)^{5-i} = 0.08146$ and $p_{\text{Car}} = p_{\text{Eng}} + p_{\text{Tires}} - p_{\text{Eng}} \cdot p_{\text{Tires}} = 0.05 + 0.08146 - 0.05 \cdot 0.08146 = 0.127387$ (Fig. 10.1).

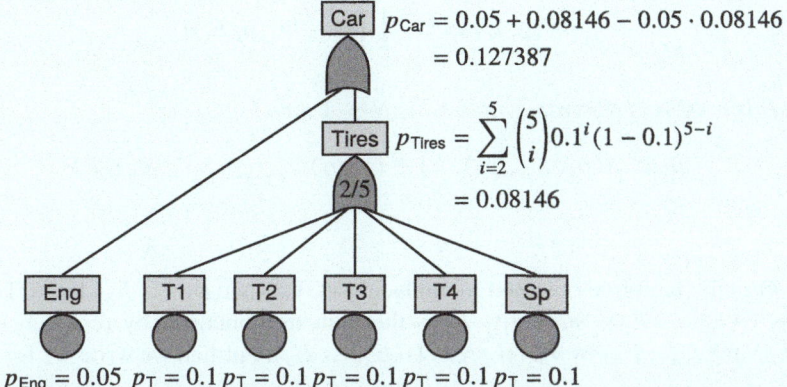

Fig. 10.1 Bottom-up computation of failure probabilities

10.3.1 Mathematical Explanation of Gate Rules

This section explains the mathematics behind the gate rules in Table 10.1. These rules can be derived from Eq. 10.1 via standard probability laws: For all stochastically independent events A, B

$$\mathbb{P}[A \wedge B] = \mathbb{P}[A]\mathbb{P}[B] \tag{10.2}$$

$$\mathbb{P}[A \vee B] = \mathbb{P}[A] + \mathbb{P}[B] - \mathbb{P}[A]\mathbb{P}[B] \tag{10.3}$$

$$\mathbb{P}[A] = \mathbb{P}[A|B] \cdot \mathbb{P}[B] + \mathbb{P}[A|\neg B] \cdot \mathbb{P}[\neg B] \tag{10.4}$$

$$\mathbb{P}[\neg A] = 1 - \mathbb{P}[A] \tag{10.5}$$

For every gate type, we consider an event E with inputs $E_1, \ldots, E_n$ and corresponding failure probabilities $p_1, \ldots, p_n$.

Mathematically, the rules for computing gate probabilities can be formulated in terms of the probabilistic structure function. An important observation is that these rules hold whenever the parent and all children are modules.

Theorem 3 *Let F be a fault tree with basic event probabilities $\mathbf{p}$. Let E be an event with children $E_1, \ldots, E_n$. If $E_1, \ldots, E_n$ are all modules in the fault tree, then we have that* *a. If E is an AND or* VOT(N/N)*-gate, then*

$$\Phi^p(\mathbf{p}, E) = \prod_{i=1}^{n} \Phi^p(\mathbf{p}, E_i)$$

b. If E is an OR or VOT($1/N$)*-gate, then*

$$\Phi^p(\mathbf{p}, E) = 1 - \prod_{i=1}^{n}(1 - \Phi^p(\mathbf{p}, E_i))$$

c. If E is a VOT(k/N)*-gate, with k > 1, N > 1 then*

$$p_1 \Phi^p_{F_1}((p_1, \ldots, p_n), E) + (1-p_1)\Phi^p_{F_2}((p_1, \ldots, p_n), E)$$

Here, F_1 is the fault tree obtained by replacing VOT(k/N)($E_1, \ldots E_n$) by VOT($k - 1/N - 1$)($E_2, \ldots E_n$). Similarly, F_2 is the fault tree obtained by replacing[1] the VOT(k/N)($E_1, \ldots E_n$) by VOT($k/N - 1$)($E_2, \ldots E_n$). Further, we write Φ^p for Φ^p_F.

The AND-gate. Given an AND-gate E with children $E_1, \ldots, E_n$, we have

$$
\begin{aligned}
p_E &= \mathbb{P}[E_1 \text{ fails} \wedge \ldots \wedge E_n \text{ fails}] \\
&= \mathbb{P}[E_1 \text{ fails}] \cdot \mathbb{P}[E_2 \text{ fails}] \cdots \mathbb{P}[E_n \text{ fails}] \\
&= p_1 \cdot p_2 \cdots p_n
\end{aligned}
$$

Expressed in terms of the probabilistic structure function, this gives the formula in Theorem 3(a): $\Phi^p(\mathbf{p}, E) = \prod_{i=1}^{n} \Phi^p(\mathbf{p}, E_i)$.

The OR-gate. For an OR-gate E with two children E_1 and E_2 we have,

$$
\begin{aligned}
p_E &= \mathbb{P}[E_1 \text{ fails} \vee E_2 \text{ fails}] \\
&= \mathbb{P}[E_1 \text{ fails}] + \mathbb{P}[E_2 \text{ fails}] - \mathbb{P}[E_1 \text{ fails} \wedge E_2 \text{ fails}] \\
&= \mathbb{P}[E_1 \text{ fails}] + \mathbb{P}[E_2 \text{ fails}] - \mathbb{P}[E_1 \text{ fails}] \cdot \mathbb{P}[E_2 \text{ fails}] \\
&= p_{E_1} + p_{E_2} - p_{E_1} \cdot p_{E_2}
\end{aligned}
$$

The formula for OR-gates with multiple children can be derived by combining multiple OR-gates with two children, i.e., OR($E_1, E_2, \ldots, E_n$) = OR(E_1, OR(E_2, OR($E_3, \ldots, E_n$))). Alternatively, the formula of the multiple-input OR-gate can be derived via the AND-gate, via De Morgan's law from Boolean algebra:

$$E_1 \vee E_2 \vee \ldots E_n = \neg(\neg E_1 \wedge \neg E_2 \wedge \ldots \wedge \neg E_n)$$

---

[1] This construction technically does not produce a fault tree, since it causes a basic event to become disconnected.

Since $\mathbb{P}[\neg E] = 1 - \mathbb{P}[E]$, we obtain

$$
\begin{aligned}
p_E &= \mathbb{P}[\neg(\neg E_1 \wedge \neg E_2 \wedge \ldots \neg E_n)] \\
&= 1 - \mathbb{P}[\neg E_1]\mathbb{P}[\neg E_2] \cdots \mathbb{P}[\neg E_n] \\
&= 1 - (1 - \mathbb{P}[E_1])(1 - \mathbb{P}[E_2]) \cdots (1 - \mathbb{P}[E_n]) \\
p_E &= 1 - (1 - p_1) \cdot (1 - p_2) \cdots (1 - p_n)
\end{aligned}
$$

Expressed in terms of the probabilistic structure function, this gives the formula in Theorem 3(b): $\Phi^p(\mathbf{p}, E) = 1 - \prod_{i=1}^{n}(1 - \Phi^p(\mathbf{p}, E_i))$.

VOT-gate. The formula for the voting gate is more complex. First, we note that the probabilities cannot be obtained by rewriting the voting gate as a combination of AND- and OR-gates. For example, $VOT(2/3)(E_1, E_2, E_3) = OR(AND(E_1, E_2),$ $AND(E_1, E_3), AND(E_2, E_3))$. We see that the rewriting introduces duplicate events, making the fault tree no longer tree-shaped, so that the bottom-up algorithm cannot be used.

Instead, we express the failure probability of the $VOT(k/N)$ gate in terms of voting gates with fewer inputs. We distinguish two cases, based on whether event E_1 fails: If E_1 fails, then we need only $k - 1$ failures among $E_2, \ldots, E_N$. Thus we obtain

$$
VOT(k/N)(E_1, \ldots, E_N) = VOT(k-1/N-1)(E_2, \ldots, E_N) \tag{10.6}
$$

If, on the other hand, E_1 does not fail, then we still need k failures among $E_2, \ldots, E_N$:

$$
VOT(k/N)(E_1, \ldots, E_N) = VOT(k/N-1)(E_2, \ldots, E_N) \tag{10.7}
$$

Next, we plug in Eqs. 10.6 and 10.7 into the formula for probabilistic case distinction. The latter states that for all events A, B we have

$$
\mathbb{P}[A] = \mathbb{P}[A|B] \cdot \mathbb{P}[B] + \mathbb{P}[A|\neg B] \cdot \mathbb{P}[\neg B]
$$

For event A we take "$VOT(k/N)(E_1, \ldots, E_N)$ fails," and for event B we take "E_1 fails."

$\mathbb{P}[VOT(k/N)(E_1, \ldots, E_N) \text{ fails}] =$
$\mathbb{P}[VOT(k/N)(E_1, \ldots, E_N) \text{ fails}|E_1 \text{ fails}] \cdot \mathbb{P}[E_1 \text{ fails}]$
$\quad + \mathbb{P}[VOT(k/N)(E_1, \ldots, E_N) \text{ fails}|E_1 \text{ does not fail}] \cdot \mathbb{P}[E_1 \text{ does not fail}] =$
$\mathbb{P}[VOT(k-1/N-1)(E_2, \ldots, E_N) \text{ fails}] \cdot p_1$
$\quad + \mathbb{P}[VOT(k/N-1)(E_2, \ldots, E_N) \text{ fails}] \cdot (1 - p_1)$

Expressed in terms of the probabilistic structure function, this gives the formula in Theorem 3(c): $p_1 \Phi^p_{F_1}((p_1, \ldots, p_n), E) + (1-p_1)\Phi^p_{F_2}((p_1, \ldots, p_n), E)$.

Voting Gate with Equal Failure Probabilities. The voting gate is often used to model redundancy, for example, in a system with 6 pumps, where at least 3 need to work for the system to be operational. This system is modeled by a 4/6 voting gate. If all pumps are identical, they are often assigned the same failure probability

p. In that case, the failure behavior of the voting gate can be described by a binomial distribution, especially through its cumulative distribution function.

For N independent components with failure probability p, let Y be the number of failed components. Then $Y \sim \text{Bin}(p, N)$ has a binomial distribution with parameters p and N. That is, the probability that exactly k components fail is given by

$$\mathbb{P}[Y = k] = \binom{N}{k} \cdot p^k (1 - p)^{N-k}$$

Hence, the probability that at least k identical components fail is given by the cumulative distribution function

$$\mathbb{P}[Y \geq k] = \sum_{i=k}^{N} \binom{N}{i} \cdot p^i (1 - p)^{N-i}$$

Thus, the failure probability for VOT(k/N)-gate with k identical and independent inputs that each fail with probability p equals

$$\mathbb{P}[\text{VOT}(k/N)(E_1, \ldots, E_N)] = \sum_{i=k}^{N} \binom{N}{i} \cdot p^i (1 - p)^{N-i}$$

There are no elementary solutions for this formula that avoid using summation. Bounds exist when k is small compared to N, but this is typically not the case for redundant systems.

10.3.2 Why the Bottom-Up Method Does Not Work for DAGs

The rules for computing gate probabilities from Table 10.1 critically rely on the independence of the branches. Therefore, the bottom-up method does not work for DAG-shaped fault trees. The problem is illustrated in Fig. 10.2, showing that events occurring on multiple branches of the tree get accounted for multiple times.

10.4 Calculating Failure Probabilities in DAG-Shaped Fault Trees

As illustrated in Fig. 10.2, the bottom-up method for computing failure probabilities only works for fault trees that are proper trees. For DAG-shaped fault trees, different methods are needed.

- We can approximate the top event probability by minimal cut sets, with algorithms explained in Chap. 8.

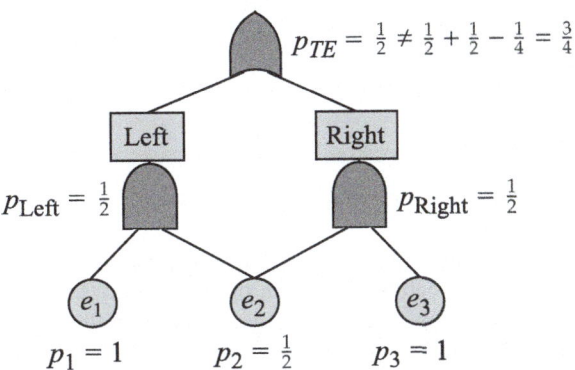

Fig. 10.2 Example showing that the bottom-up method does not work in DAG-shaped fault trees. Take $p_1 = p_3 = 1$ and $p_2 = \frac{1}{2}$. Then $p_{Top} = \frac{1}{2}$, as only e_2 has to probabilistically fail for the top event to occur. When using the gate rules from Table 10.1, the probabilities p_{Left} and p_{Right} are assigned the correct values, but the probability for the top event is not. The problem is that p_2 gets accounted for in both branches

- The most common methods for an exact calculation of the top event probability are based on binary decision diagrams (BDD). This method is explained in the next section.
- One can also use conditional probabilities to compute the top probability. This method is less applied in practice, but provides a good perspective on the nature of the problem. This method is explained in Sect. 10.6.

10.5   Calculating the Top Probability via BDDs

Recall from Sect. 3.2 that a BDD is a common representation of the structure function of a fault tree. In Sect. 9.2, we explained how BDDs can be used to encode the cuts sets of a fault tree. In this section, we show how to decorate these BDDs to obtain the top probability.

Binary decision diagrams were introduced in Chap. 3, where Sect. 3.2 presented their definition, and Sect. 3.2.1 explained how to obtain a BDD from a fault tree. Material presented in those sections is preliminary to understand this section.

Given a BDD encoding of a fault tree, computing the top probability is conceptually simple. It consists of the following two steps.

1. First, one decorates each edge in the BDD with a probability. Let v be a BDD node labeled with basic event e.

- The (solid) edge from v to its right child v_R gets assigned probability $\mathbb{P}[e]$.
- The (dashed) edge from a BDD node v to its left child v_L gets assigned probability $1 - \mathbb{P}[e]$.

$$p^{\text{edge}}(v, v_R) = \mathbb{P}[e]$$
$$p^{\text{edge}}(v, v_L) = 1 - \mathbb{P}[e]$$

Indeed, taking the solid edge in the BDD corresponds to basic event e failing, which happens with probability $\mathbb{P}[e]$; taking the dashed (left) edge corresponds to basic event e not failing, which happens with probability $1 - \mathbb{P}[e]$.

2. Next, one adds all path probabilities from the BDD's root node R to any leaf labeled by 1, where each path probability is obtained by multiplying all probabilities along the edges. We set:

$$p^{\text{path}}(\pi) = \prod_i p^{\text{edge}}(\pi_i)$$
$$p(R) = \sum_{\pi \text{ is 1-path}} p^{\text{path}}(\pi)$$

Here, π is a path in the BDD, i.e., a sequence $(v_1, v_2)(v_2, v_3) \ldots (v_{n-1}, v_n)$ such that v_{i+1} is a (left or right) child of v_i. A 1-path is a path that starts in the root R and ends in a node labeled 1. The method is exemplified in Fig. 10.3 and Table 10.2.

Then the following property holds

Theorem 4 *Let R be the root of the BDD encoding of fault tree F and let Top be the top level event of F. Then*

$$\mathbb{P}[Top] = p(R)$$

To see that this method is correct, we observe the following:

- Every path in the BDD leading to a 1-leaf corresponds to a set of status vectors that make the FT fail.
- The probability for that set of status vectors to occur is obtained by multiplying their probabilities.
- Every path in the BDD corresponds to a disjoint set of status vectors. Hence, their probabilities can be added.

Note that this computation is correct, irrespective of the chosen BDD ordering.

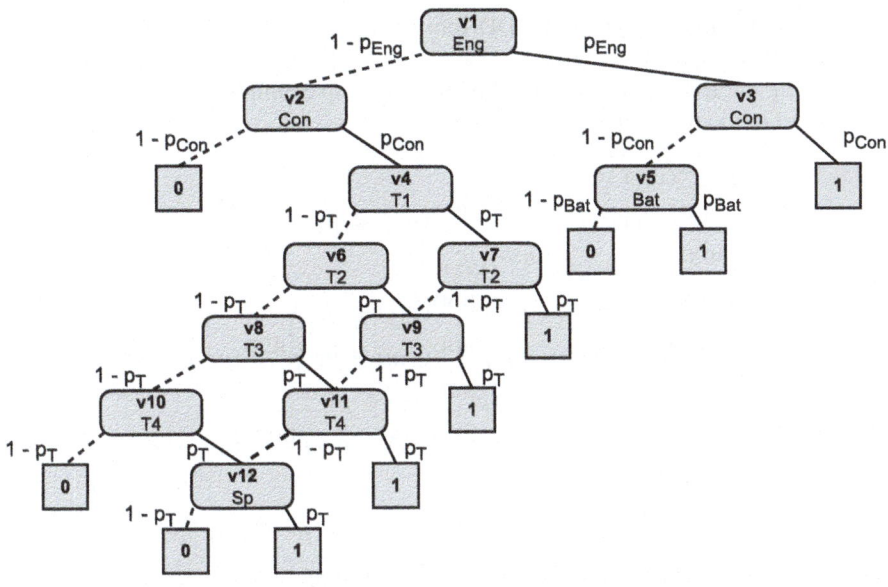

Fig. 10.3 BDD for the road trip fault tree annotated with failure probabilities

Table 10.2 1-Paths of the BDD in Fig. 10.3 and their associated probabilities

BDD path	Probability
Eng, Con	$p_{Eng} \cdot p_{Con}$
Eng, ¬Con, Bat	$p_{Eng}(1 - p_{Con})p_{Bat}$
¬Eng, Con, T1, T2	$(1 - p_{Eng})p_{Con}\, p_T\, p_T$
¬Eng, Con, ¬T1, T2, T3	$(1 - p_{Eng})p_{Con}(1 - p_T)p_T\, p_T$
¬Eng, Con, ¬T1, T2, ¬T3, T4	$(1 - p_{Eng})p_{Con}(1 - p_T)p_T(1 - p_T)p_T$
¬Eng, Con, ¬T1, T2, ¬T3, ¬T4,Sp	$(1 - p_{Eng})p_{Con}(1 - p_T)p_T(1 - p_T)(1 - p_T)p_T$
¬Eng, Con, ¬T1, ¬T2, T3, T4	$(1 - p_{Eng})p_{Con}(1 - p_T)(1 - p_T)p_T\, p_T$
¬Eng, Con, ¬T1, ¬T2, ¬T3, T4, Sp	$(1 - p_{Eng})p_{Con}(1 - p_T)(1 - p_T)(1 - p_T)p_T\, p_T$

10.5.1 Efficient Computation

The method described above is still inefficient, since it ranges over all paths in the BDD. However, the sum of path probabilities can be computed efficiently with a bottom-up procedure over the BDD. The procedure avoids duplicating work if there are overlapping paths in the BDD.

$$
p^{BDD}(v) = \begin{cases} (1 - p_e)p^{BDD}(v_L) + p_e p^{BDD}(v_R) & \text{if } v \text{ is an internal node} \\ 0 & \text{if } v \text{ is a leaf node labeled 0} \\ 1 & \text{if } v \text{ is a leaf node labeled 1} \end{cases}
$$

Then the probability assigned to BDD root node equals the top event probability, i.e., $p^{\text{BDD}}(R) = \mathbb{P}[Top]$.

Applying the BDD-bottom method on the road trip example yields the following equations: On the road trip example, we have

$$p^{\text{BDD}}(v_{12}) = p_{\text{T}} \cdot 1$$
$$p^{\text{BDD}}(v_{11}) = (1 - p_{\text{T}})p^{\text{BDD}}(v_{12}) + p_{\text{T}} \cdot 1$$
$$p^{\text{BDD}}(v_{10}) = p_{\text{T}}p^{\text{BDD}}(v_{12})$$
$$p^{\text{BDD}}(v_{9}) = (1 - p_{\text{T}})p^{\text{BDD}}(v_{11}) + p_{\text{T}} \cdot 1$$
$$p^{\text{BDD}}(v_{8}) = (1 - p_{\text{T}})p^{\text{BDD}}(v_{10}) + p_{\text{T}} \cdot p^{\text{BDD}}(v_{11})$$
$$p^{\text{BDD}}(v_{7}) = (1 - p_{\text{T}})p^{\text{BDD}}(v_{9}) + p_{\text{T}} \cdot 1$$
$$p^{\text{BDD}}(v_{6}) = (1 - p_{\text{T}})p^{\text{BDD}}(v_{8}) + p_{\text{T}} \cdot p^{\text{BDD}}(v_{9})$$
$$p^{\text{BDD}}(v_{5}) = p_{\text{Bat}} \cdot 1$$
$$p^{\text{BDD}}(v_{4}) = (1 - p_{\text{T}})p^{\text{BDD}}(v_{6}) + p_{\text{T}} \cdot p^{\text{BDD}}(v_{7})$$
$$p^{\text{BDD}}(v_{3}) = (1 - p_{\text{Con}})p^{\text{BDD}}(v_{5}) + p_{\text{Con}} \cdot 1$$
$$p^{\text{BDD}}(v_{2}) = p_{\text{Con}} \cdot p^{\text{BDD}}(v_{4})$$
$$p^{\text{BDD}}(v_{1}) = (1 - p_{\text{Eng}})p^{\text{BDD}}(v_{2}) + p_{\text{Eng}} \cdot p^{\text{BDD}}(v_{3})$$

10.6 Calculating the Top Probability via Conditional Probabilities

A second method to compute the probabilities for the top and intermediate events is through the laws of conditional probability. This method extends the bottom-up method by keeping track of shared events, so they can be correctly accounted for. This method is not often used in practice, but gives a useful perspective on the nature of fault tree quantification. We illustrate the method on the road trip example.

$$p_{\text{Power}} = p_{\text{Bat}} \cdot p_{\text{Eng}}$$
$$p_{\text{Phone}} = p_{\text{Con}} + p_{\text{Power}} - p_{\text{Con}} \cdot p_{\text{Power}}$$
$$= p_{\text{Con}} + p_{\text{Bat}} \cdot p_{\text{Eng}} - p_{\text{Con}} \cdot p_{\text{Bat}} \cdot p_{\text{Eng}} \qquad (10.8)$$

To obtain p_{Tires} we consider the complement probability that the tire system does not fail within the given time horizon. This happens if either five or four tires do not fail. The probability that all tires are up, i.e., none of them fails is given by $(1 - p_{\text{T}})^5$. Further, the probability that one specific tire i has failed, while all other four are up is given by $p_{\text{T}}(1 - p_{\text{T}})^4$. Since there are five ways of choosing the specific tire, we obtain $5 \cdot p_{\text{T}}(1 - p_{\text{T}})^4$. Now, the probability that the tire system fails is given by

$$p_{\text{Tires}} = 1 - (1 - p_{\text{T}})^5 - 5 \cdot p_{\text{T}}(1 - p_{\text{T}})^4$$

Since the Car is an OR-gate, we obtain

$$p_{Car} = 1 - (1 - p_{Eng}) \cdot \left((1 - p_T)^5 - 5 \cdot p_T (1 - p_T)^4\right) \tag{10.9}$$

Finally, to compute $p_{Road\ trip}$, we need to take into account the fact that the subtrees are not independent, since they share the Eng gate. We use Bayes law

$$\mathbb{P}[A] = \mathbb{P}[A|B] \cdot \mathbb{P}[B] + \mathbb{P}[A|\neg B] \cdot \mathbb{P}[\neg B]$$

stating that the probability for an event A can be computed by distinguishing two cases: either B happens or B does not happen. However, these cases must be weighted by their respective probabilities. Thus, the probability for A to happen equals the probability for A to happen given that event B happens, weighted by probability for B to happen, plus the probability for A to happen given that B does not happen, weighted by probability for B to not happen.

To obtain $p_{Road\ trip}$, we consider two cases: either Eng has failed, or it has not. Thus, we obtain

$$\mathbb{P}[\text{Road trip fails}] = \mathbb{P}[\text{Road trip fails}|\text{Eng fails}] \cdot \mathbb{P}[\text{Eng fails}] +$$
$$\mathbb{P}[\text{Road trip fails}|\text{Eng does not fail}] \cdot \mathbb{P}[\text{Eng does not fail}]$$

Further, we use

$$\mathbb{P}[\text{Road trip fails}|\text{Eng fails}] = \mathbb{P}[\text{Phone fails} \wedge \text{Car fails}|\text{Eng fails}]$$
$$= \mathbb{P}[\text{Phone fails}|\text{Eng fails}] \cdot \mathbb{P}[\text{Car fails}|\text{Eng fails}]$$

The latter equality holds because, under the assumption that the engine has failed, the failure of the phone and car are independent. Hence, we can multiply their probabilities. Finally, to obtain $\mathbb{P}[\text{Phone fails}|\text{Eng fails}]$, use Eq. 10.8, and plug in $p_{Eng} = 1$, since the engine has failed, yielding

$$\mathbb{P}[\text{Phone fails}|\text{Eng fails}] = p_{Con} + p_{Bat} \cdot p_{Eng} - p_{Con} \cdot p_{Bat} \cdot p_{Eng}$$
$$= p_{Con} + p_{Bat} - p_{Con} \cdot p_{Bat}$$

Similarly, to obtain $\mathbb{P}[\text{Car fails}|\text{Eng fails}]$, we use Eq. 10.8, but with $p_{Eng} = 0$, since the engine does not fail, yielding

$$\mathbb{P}[\text{Phone fails}|\text{Eng does not fail}] = p_{Con} + p_{Bat} \cdot p_{Eng} - p_{Con} \cdot p_{Bat} \cdot p_{Eng}$$
$$= p_{Con}$$

In a similar way, we obtain from Eq. 10.9 that

$$\mathbb{P}[\text{Car fails}|\text{Eng fails}] = 1$$
$$\mathbb{P}[\text{Car fails}|\text{Eng does not fail}] = 1 - (1 - p_T)^5 - 5 \cdot p_T (1 - p_T)^4$$

Together this yields the following equation for the top level probability.

$$\mathbb{P}[\text{Road trip fails}] = \left(p_{\mathsf{Con}} + p_{\mathsf{Bat}} - p_{\mathsf{Con}} p_{\mathsf{Bat}}\right) p_{\mathsf{Eng}} + p_{\mathsf{Con}} \overline{p_{\mathsf{Eng}}} \left(1 - \overline{p_{\mathsf{T}}}^5 - 5 p_{\mathsf{T}} \overline{p_{\mathsf{T}}}^4\right)$$

(10.10)

10.7 Mathematical Formulation

We end this chapter with a formalization of the probabilistic calculations in terms of random variables. Chapter 11 extends this formulation to the time-dependent case, using continuous random variables.

As stated, each basic event e gets equipped with a failure probability p_e. Mathematically, we assign to basic event e a random variable X_e with a binary (a.k.a. Bernoulli) distribution with parameter p_e. That is, $X_e \sim \text{Binary}(p_e)$ is given by

$$\mathbb{P}[X_e = 1] = p_e$$
$$\mathbb{P}[X_e = 0] = 1 - p_e$$

Furthermore, we assume that all basic events are independent, that is, the set of random variables $\{X_{e_1}, \ldots, X_{e_n}\}$ is mutually independent.

Since the structure function Φ is a Boolean function, applying the structure function to the basic event random variables $X_{e_1}, \ldots, X_{e_n}$ yields a random variable Y. Then Y has a binary probability distribution as well, and its parameter yields top failure probability. Formally, $Y = \Phi(X_{e_1}, \ldots, X_{e_n}) \sim \text{Binary}(p_{Top})$ and we set

$$p_{Top} = \mathbb{P}[\Phi(X_{e_1}, \ldots, X_{e_n}) = 1]$$

(10.11)

The same holds for intermediate events. If E is an intermediate event, then $\Phi(X_{e_1}, \ldots, X_{e_n}, E)$ is a random variable with a binary distribution and we set $p_E = \mathbb{P}[\Phi(X_{e_1}, \ldots, X_{e_n}, E) = 1]$. In this way, $X_E = 1$ is equivalent to saying "Event E fails," and $X_E = 0$ is equivalent to "Event E does not fail" or "E is operational." Now the following result follows.

Theorem 5 *For an intermediate event E with children $E_1, \ldots, E_n$, we have*

$$p_E = $$

$$\begin{cases} \mathbb{P}[E_1 \text{ fails} \wedge \ldots \wedge E_n \text{ fails}] & E \text{ is an AND-gate} \\ \mathbb{P}[E_1 \text{ fails} \vee \ldots \vee E_n \text{ fails}] & E \text{ is an OR-gate} \\ \mathbb{P}[\text{There are } i_1, i_2, \ldots, i_k : E_{i_1} \text{ fails} \wedge \ldots \wedge E_{i_k} \text{ fails}] & E \text{ is a VOT}(k/N)\text{-gate} \end{cases}$$

Dependability Metrics for Fault Trees Without Repair

11

Time-dependent fault tree analysis studies the failure behavior of a system over time. To do so, several *dependability metrics* have been defined to evaluate how the failure probabilities evolve over time. We focus on the most common ones.

- The *availability* is the probability that the system is operational. Several variants exist: the point availability is the probability that the system is operational at a given time point t. The long-run average availability is the probability that the system is available in the long run, i.e., when time goes to infinity.
- The *reliability* is the probability that the system has not failed before a given time point t, often called the mission time.
- The *mean time to failure* is the average time it takes for the system to fail.

These metrics act as key performance indicators and play a crucial role in risk-based decision-making, e.g., for comparing design alternatives or assessing compliance with standards. As such, they apply to dependability analysis in general, not only to fault trees. In the context of fault trees, these metrics can be studied for all events, i.e., basic, intermediate and the top event. When considering the top event, we use the terms *system* reliability and *system* availability. We study two setups.

- The current chapter discusses systems without repair, while the next chapter elaborates on systems with repair. In both cases, each basic event is assigned a failure time distribution, i.e., a continuous probability distribution function that describes the probability that event e fails before a given time t.
- The next chapter considers systems that can be repaired, allowing one to study the effect of testing, maintenance and repairs. Here, each basic event is assigned, in addition to the failure distribution, also a repair time distribution. We study two repair models: with observable and unobservable failures.

© The Author(s), under exclusive license to Springer Nature Switzerland AG 2026   129
M. Stoelinga et al., *Concise Guide to Fault Tree Analysis*, Computer Science Foundations and Applied Logic, https://doi.org/10.1007/978-3-031-78287-9_11

It turns out that analyzing the availability in the time-dependent case closely mirrors the approach used for the time-independent case, discussed in the previous chapter. This holds both for the point and the long-run average availability. The key lies in studying the time-dependent variants $p_i(t)$ of the point probabilities p_i. The exact shape of $p_i(t)$ depends on the reliability model, i.e., whether or not repairs are considered, and, if so, whether failures are observable. Thus, the availability metrics can be computed by reusing methods from the previous chapter, by simply substituting p_i by $p_i(t)$. As we will show, this substitution does not work for the reliability. Specifically, in the repairable case, the reliability cannot be obtained via the point probabilistic model.

11.1 Overview

We start with an overview of the main approaches, an overview is shown in Table 11.1.

11.1.1 Systems Without Repair

Availability. For systems that cannot be repaired, one typically studies the *point availability,* i.e., the probability that the system is operational at a specific point in time. The point availability $\mathsf{Availability}^{\mathsf{pt}}(t)$ can be viewed as a time-dependent variant of the point probabilistic model from Chap. 10, where each failure probability depends on time. Since these time dependencies smoothly propagate over the structure function, the point availability can be studied with techniques similar to the point probabilistic model.

Reliability. Another metric commonly studied is the *reliability,* i.e., the probability that the system has not failed before a given time t, denoted $\mathsf{Reliability}(t)$. For example, one can study the reliability for a mission to the moon that lasts for 9 months. For irreparable systems, the reliability for mission time t equals the point availability at t. While both metrics have different use cases, they are computed with the same methods and work for a large class of failure distributions.

11.1.2 Systems with Repair

We first consider the case where all failures are observable, and are repaired immediately using a repair distribution. It is assumed that repair is perfect and does not affect future failure or repair distributions.

Availability. The point availability can be treated in exactly the same way as for systems without repair, which, as we saw, is also similar to the point probabilistic model.

Table 11.1 Analysis of time-dependent dependability metrics

	No repair	Repairs
Availability$^{pt}(t)$	Point probabilistic model	Point probabilistic model
Reliability(t)	$=$ Availability$^{pt}(t)$	CTMC/semi-Markov process
Availability^{lr}	$= 0$	Point probabilistic model

It is also common to study the *long run average availability*, i.e., the behavior when the system is in its steady state, denoted Availability^{lr}. (Note that the long run average availability is 0 for irreparable systems, because every system will fail eventually.)

In the case of observable failures, the long run availability can be calculated as a fraction MTTF/(MTTF + MTTR), where MTTF is the mean time to failure and MTTR the mean time to repair.

Second, we consider the case where components are subject to unobservable failures. These are usually tested periodically, and if a failure is found, the component is repaired. The definition of long run average availability is slightly more involved in this case.

However, both for observable and unobservable failures, computation for the long run average availability is similar to the point probabilistic model. Moreover, these techniques work for a wide range of failure and repair distributions.

11.2   Beyond Reliability and Availability

Numerous other reliability metrics exist. We briefly touch on the mean time to failure and mean time to repair. Several of these measures require analysis of the CTMC or the semi-Markov process underlying the fault tree.

Reliability. The *system reliability* considers the failure behavior of a system before a given mission time. Computing the reliability for repairable systems is complex, since several component failures and repairs may occur before the system fails. Therefore, reliability calculations involve the semi-Markov process underlying the fault tree.

This semi-Markov process is a stochastic transition system that describes the complete failure behavior of the fault tree. The states of this transition system are the status vectors of the fault tree, and its transitions describe the joint failure and repair behavior of all basic events, for each status vector. Since a fault tree with n basic events has 2^n status vectors, the transition system has 2^n states. Nevertheless, several special cases are feasible for small and medium sized fault trees. In particular, if all failures and repairs are exponential, then the transition system is a continuous time Markov chain, for which numerous fast analysis techniques exist. In other cases, dedicated techniques are needed, or Monte Carlo simulation can be used.

11.3   Examples: Point Availability and Reliability

11.3.1   Time-Dependent Failures in the Road Trip Example

Analysis of fault trees with time-dependent probabilities and no repairs is not fundamentally different from the point probabilistic model. The key is to study a time-dependent version of all failure probabilities. That is, for each basic event, we use a time-dependent failure probability $p_i(t)$, rather than a point probability p_i.

We illustrate these principles on the road trip example. Several examples in this chapter use the exponential probability distribution. We assume the reader is already familiar with the concepts covered in Sect. 3.4.

Example 23 (Road trip example with time-dependent failures.)

Section 10.6 showed that the time-independent probability for the road trip to fail equals

$$p_{\text{Top}} = (p_1 + p_2 - p_1 p_2)\, p_3 + p_1 \overline{p_3} \left(1 - \overline{p_4}^5 - 5 p_4 \overline{p_4}^4 \right)$$

Now, for each time t, we consider $p_i(t)$ to be the probability that basic event e_i is failed at time t. Then

$$p_{\text{Top}}(t) = (p_1(t) + p_2(t) - p_1(t) p_2(t))\, p_3(t) +$$
$$p_1(t)\overline{p_3(t)} \left(1 - \overline{p_4(t)}^5 - 5 p_4(t)\overline{p_4(t)}^4 \right)$$

In fact, $p_i(t)$ is the point availability of e_i at time t, and $p_{\text{Top}}(t)$ is the point availability at time t. Below, we show that for irreparable systems, the point availability equals the reliability. Therefore,

$$\text{Reliability}(t) = \text{Availability}^{\text{pt}}(t) = p_{\text{Top}}(t)$$

Now assume that each basic event e is exponentially distributed with parameter λ_i, that is, $p_i(t) = 1 - e^{-\lambda_i t}$. The exponential distribution is elaborated in Sect. 3.4. Substitution yields

$$p_{\text{Top}}(t) = \left(1 - e^{-\lambda_1 t} + 1 - e^{-\lambda_2 t} - (1 - e^{-\lambda_1 t})(1 - e^{-\lambda_2 t})\right)\left(1 - e^{-\lambda_3 t}\right) +$$
$$(1 - e^{-\lambda_1 t})e^{-\lambda_3 t}\left(1 - (e^{-\lambda_4 t})^5 + 5(1 - e^{-\lambda_4 t})(e^{-\lambda_4 t})^4\right)$$

Analysis at fixed time points. It is important to realize that, when we study the reliability or availability at a fixed point in time, then we are back in the point probabilistic model. That is, we are just working with numbers, not with functions. For example, for a fixed mission time of $t = 10$ days, the basic event probability $p_e(10)$ is a constant. Consider the road trip example for $t = 10$ and the following values for the failure rates λ_i:

- $\lambda_1 = 4$, so $p_1(0.5) = 1 - e^{-2} \approx 0.865$
- $\lambda_2 = 5$, so $p_2(0.5) = 1 - e^{-2.5} \approx 0.918$
- $\lambda_3 = 20$, so $p_3(0.5) = 1 - e^{-10} \approx 0.999955$
- $\lambda_4 = 10$, so $p_4(0.5) = 1 - e^{-5} \approx 0.993$

Substituting these numbers into p_{Top} in Example 23 yields

$$
\begin{aligned}
p_{\text{Top}} &= (p_1 + p_2 - p_1 p_2)p_3 + p_1 \overline{p_3}\left(1 - \overline{p_4}^5 - 5 p_4 \overline{p_4}^4\right) \\
&= (0.865 + 0.918 - 0.865 \cdot 0.918)0.999955 \\
&\quad + 0.865 \cdot 0.000045\left(1 - 0.006^5 - 5 \cdot 0.993 \cdot 0.006^4\right) \\
&= 0.988891 \cdot 0.999955 + 0.0000393 \cdot 0.9999999897 \\
&= 0.988846 + 0.0000393 \\
&= 0.989
\end{aligned}
$$

This example also shows that it is important to take into account the precision of the real numbers in the computation. Next, we illustrate that the substitution method also works for other probability distributions. We choose a lognormal distribution, but the method works for any distribution.

Example 24 (Road trip example with time-dependent failures)

The exact same method works for other probability distributions. Suppose that the component failures follow a lognormal distribution, where the parameters for e_i are μ_i and σ_i. That is

$$
p_i(t) = \Phi\left(\frac{\ln(t) - \mu_i}{\sigma_i}\right)
$$

where Φ is the cumulative distribution function of the standard normal distribution. Again, we can simply substitute the formula $p_i(t)$ in p_{Top}.

$$
\begin{aligned}
p_{\text{Top}}(t) = {}& \text{Reliability}(t) = \text{Availability}^{\text{Pt}}(t) \\
= {}& \left(\Phi\left(\frac{\ln(t) - \mu_1}{\sigma_1}\right) + \Phi\left(\frac{\ln(t) - \mu_2}{\sigma_2}\right) - \Phi\left(\frac{\ln(t) - \mu_1}{\sigma_1}\right)\Phi\left(\frac{\ln(t) - \mu_2}{\sigma_2}\right)\right) \cdot \\
& \Phi\left(\frac{\ln(t) - \mu_3}{\sigma_3}\right) + \Phi\left(\frac{\ln(t) - \mu_1}{\sigma_1}\right)\left(1 - \Phi\left(\frac{\ln(t) - \mu_3}{\sigma_3}\right)\right) \cdot \\
& \left(1 - \left(1 - \Phi\left(\frac{\ln(t) - \mu_4}{\sigma_4}\right)\right)^5 - 5\Phi\left(\frac{\ln(t) - \mu_4}{\sigma_4}\right)\left(1 - \Phi\left(\frac{\ln(t) - \mu_4}{\sigma_4}\right)\right)^4\right)
\end{aligned}
$$

11.4 The Time-Dependent Probabilistic Model Without Repair

This section provides the probabilistic model needed to define the dependability metrics for time-dependent systems. We define the status function and failure time as continuous random variables. The dependability metrics are then defined in terms of these random variables.

11.4.1 Status Function

Status function for basic events. The starting point for time-dependent analysis is to make the system status a time-dependent function [1,2]. Here, different time concepts may be used, such as calendar time, the number of kilometers/miles driven by a car, the number of operating hours for a plant, or the rotations of a bearing. For each event E in a fault tree, we describe its failure behavior over time by the function X_E, called the *status function*, also known as the state or status variable. That is, $X_E(t)$ denotes the status of event E at time t:

$$X_E(t) = \begin{cases} 1 & \text{Event } E \text{ is failed at time } t \\ 0 & \text{Event } E \text{ is operational at time } t \end{cases}$$

The status function is a random variable, so $\mathbb{P}[X_E(t) = 1]$ is the probability that event E is failed at time t, and $\mathbb{P}[X_E(t) = 0]$ is the probability that event E is operational at time t, see Fig. 11.1. For repairable systems, X_E can switch between 0 and 1 multiple times; for unrepairable systems, it will only switch to 1 a single time. By setting

$$p_E(t) = \mathbb{P}[X_E(t) = 1]$$

we see that the status function is a time-dependent variant of the event probabilities p_E from the point probabilistic model.

Status function for intermediate and top events. Given the basic event status functions $X_1, \ldots, X_n$, the status functions for intermediate and top events can be obtained via the structure function Φ of the fault tree in question. Then the status function for an intermediate event E and top event Top are given by

$$X_E = \Phi_F((X_1, \ldots, X_n), E)$$
$$X_{\text{Top}} = \Phi_F(X_1, \ldots, X_n)$$

Note that $\Phi_F((X_1, \ldots, X_n), E)$ and $\Phi_F(X_1, \ldots, X_n)$ are random variables in continuous time: applying any function f to a random variable X yields a random variable $f(X)$. Therefore, we can ask for the probability that such a random variable equals 1. The definitions of X_E and X_{Top} yield the following equations.

$$\mathbb{P}[X_E(t) = 1] = \mathbb{P}[\Phi_F((X_1(t), \ldots, X_n(t)), E) = 1] \qquad (11.1)$$
$$\mathbb{P}[X_{\text{Top}}(t) = 1] = \mathbb{P}[\Phi_F(X_1(t), \ldots, X_n(t)) = 1] \qquad (11.2)$$

Note that, if we consider X_E at a fixed time point t_0, then $\mathbb{P}[X_E(t_0) = 1] = p_E$ yields the point probabilistic model from Sect. 10.7.

11.4.2   Failure Times and Failure Time Distributions

Failure times. From X_E, we derive T_E, the *failure time* of E. T_e is the point in time when E fails for the first time, see Fig. 11.1. Thus, the failure time T_E is defined as

$$T_E = \min\{t \in [0, \infty) \mid X_E(t) = 1\}$$

Since X_E is a random variable, so is T_E. We write F_E for its distribution function, yielding the probability that E fails before time t, see Fig. 11.1. We define

$$F_E(t) = \mathbb{P}[\text{Event } E \text{ fails before time } t] = \mathbb{P}[T_E < t]$$

Commonly, F_E is called the *failure (time) distribution* of E. Common failure time distributions include the exponential distribution (elaborated in Sect. 3.4), the Weibull and lognormal distributions.

Remark 1 Mathematically, F_E is defined as the cumulative distribution function (CDF) of T_E. Therefore, F_E is increasing, i.e., $t' > t$ implies $F_E(t') > F_E(t)$. Intuitively, this is clear: the longer the time, the more opportunities exist for E to fail. Moreover, $\lim_{t \to \infty} F_E(t) = 1$, that is, every event eventually fails. We assume that the probability to fail at exactly time t is usually zero, i.e., $\mathbb{P}[T_E = t] = 0$. Finally, $\mathbb{P}[T_E < t] = \mathbb{P}[T_E \le t]$, $\mathbb{P}[T_E > t] = 1 - \mathbb{P}[T_E < t]$ and $\mathbb{P}[t_1 < T_E < t_2] = F_E(t_2) - F_E(t_1)$.

Often, one is interested not only in the failure probabilities over time, but also when these failure probabilities increase. The probability density function f_E is the derivative of F_E, indicating how fast F_E changes over time (Fig. 11.2).

$$f_E(x) = F'_E(t) \text{ and } F_E(t) = \int_0^t f_E(x)\mathrm{d}x$$

The above definitions apply to the basic, intermediate and top events. As before, we assign failure time distributions to all basic events e and calculate failure time

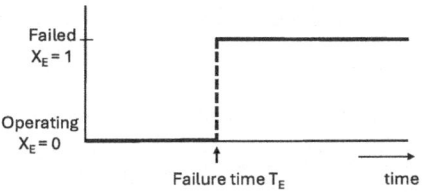
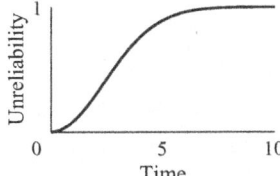

Fig. 11.1 Status variable X_E and time to failure F_E

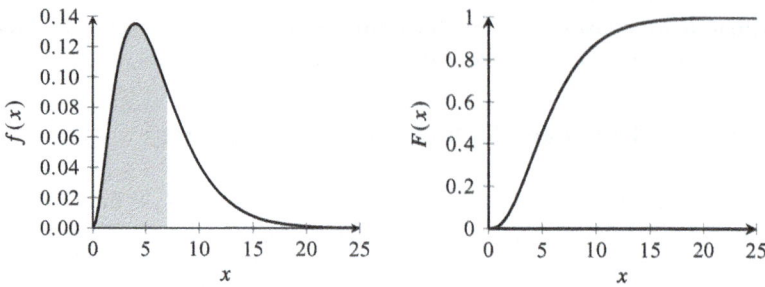

Fig. 11.2 Probability density function and cumulative distribution function [4]

Table 11.2 Overview of notation for time-dependent failure behavior. RV = random variable, CDF = cumulative distribution function

Notation	Term	Meaning
$X_E(t)$	Status function	RV denoting status of event E at time t
$T_E = \min\{t\|X_E(t) = 1\}$	Failure time	RV denoting time of first failure of E
$F_E(t) = \mathbb{P}[T_E < t]$	Failure time distribution	CDF yielding probability for E to fail before time t

distributions and other dependability metrics for other events, especially the top event.

Defining failure distributions. Mathematically, it is common to define the failure distribution F_E in terms of T_E, which is defined in terms of X_E, which is in turn defined via $X_1, \ldots, X_n$. So, if we know X_E, then we know T_E and F_E. In practice, we often obtain F_E from data, following one of the reliability models from Sect. 10.1 and then derive X_E and T_e.

Overview of notation. Table 11.2 summarizes notation used.

11.4.3 Obtaining Failure Distributions

An important question in fault tree analysis is to determine what failure distribution to use. One of the most common failure time distributions is the *exponential distribution*, which is extensively covered in Sect. 3.4. The exponential distribution assumes that the number of failures per time unit is constant, as shown in the famous bathtub curve, see Fig. 11.3.

The bathtub curve shows the number of failures over time of a component or system. Often, there are many failures at the beginning, due to birth defects such as installation failures. During the normal life, the number of failures per time unit is often constant, until at the end, where the number of failures increase again. If the

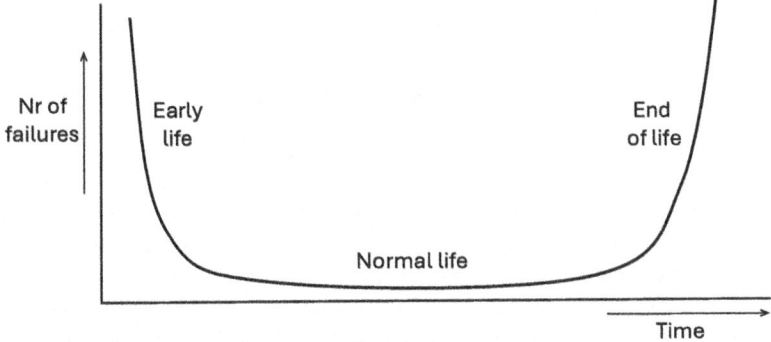

Fig. 11.3 Bathtub curve, showing number of failures over time of a component or system

number of failures during normal life is constant, then an exponential distribution provides a realistic failure model. However, if wear is an important failure cause, then the number of failures may increase during normal life, and other distributions can be useful. Common failure distributions include the Weibull, gamma, normal, lognormal, and the inverse Gaussian distributions. Choosing which distribution to use when falls outside the scope of this book. We refer to [1] for an in-depth treatment.

11.5   Dependability Metrics for Systems Without Repair

When studying time-dependent failure behavior, several metrics are of interest.

Availability. One of the most common metrics is the *system availability*, or *availability* for short. Several definitions of availability exist. Regardless whether the system is repairable, the *point availability* is relevant, which is defined as the probability that the system is operational at a given point time t. This means that the status function equals 0 at time t.

$$\text{Availability}^{\text{pt}}(t) = \mathbb{P}[\text{system is operational at time } t]$$
$$= \mathbb{P}[X_{\text{Top}}(t) = 0]$$

As shown in Eq. 11.2, the point availability can be derived from the status functions $X_1, \ldots, X_n$ of the basic events via the structure function. Thus, we obtain:

$$\text{Availability}^{\text{pt}}(t) = \mathbb{P}[X_{\text{Top}}(t) = 0]$$
$$= \mathbb{P}[\Phi_F(X_1(t), \ldots, X_n(t)) = 0]$$

As usual, the unavailability is defined as the complement of the availability, yielding

$$\text{Unavailability}^{\text{pt}}(t) = \mathbb{P}[\text{system is failed at time } t]$$
$$= \mathbb{P}[X_{\text{Top}}(t) = 1]$$
$$= 1 - \text{Availability}^{\text{pt}}(t)$$

The point availability can also be defined for other events, setting $\text{Availability}^{\text{pt}}(t) = \mathbb{P}[X_E(t) = 0]$. In fact, the probabilities $p_i(t)$ in Sect. 11.3 are point availabilities and $p_{\text{Top}}(t)$ is the point availability for the top event.

For repairable systems, it is common to consider the long-run average availability, which considers the availability when time goes to infinity. However, systems without repair will always fail eventually. Therefore their long run average availability equals 0: $\text{Availability}^{\text{lr}} = \lim_{t \to \infty} \text{Availability}^{\text{pt}}(t) = 0$.

Reliability. Another common metric is the *system reliability*, or *reliability* for short, defined as the probability that the system does not fail within its mission time t. Its complement is also common, the *(system) unreliability*, being the probability that the system does fail within time t.

$$\text{Reliability}(t) = \mathbb{P}[\text{system does not fail before time } t] = 1 - F_{\text{Top}}(t) = \mathbb{P}[T_{\text{Top}} > t]$$
$$\text{Unreliability}(t) = \mathbb{P}[\text{fails before time } t] = 1 - \text{Reliability}(t) = F_{\text{Top}}(t)$$

The reliability is also called the *survivor function*, as it asks whether the system has survived the first t time units.

Reliability equals availability. For coherent fault trees without repair, the availability and reliability coincide: The probability that the system has not failed before time t equals the probability that the system is operational at time t. This holds because the system fails only once: there are no repairs, and due to coherency, additional basic event failures cannot render a failed system operational.

We can derive this fact as follows. First we observe that for coherent systems without repair it holds that.

$$X_{\text{Top}}(t) = 1 \wedge t < t' \implies X_{\text{Top}}(t') = 1. \qquad (*)$$

Therefore,

$$\text{Reliability}(t) = 1 - F_{\text{Top}}(t)$$
$$= 1 - \mathbb{P}[T_{\text{Top}} < t]$$
$$= 1 - \mathbb{P}[\min\{t' \mid X_{\text{Top}}(t') = 1\} < t] \qquad \text{due to } (*)$$
$$= 1 - \mathbb{P}[X_{\text{Top}}(t) = 1]$$
$$= \text{Availability}^{\text{pt}}(t)$$

For repairable systems, the reliability and availability are different. The reliability only considers the first failure, whereas the availability considers subsequent failures as well.

MTTF. From the system reliability, one obtains the *mean time to failure (MTTF)*, sometimes called the *mean time to first failure* or *mean time before failure*. The MTTF is the average time that it takes for a system to fail, when starting in pristine condition. Since T_{Top} represents the failure behavior of the system, $\mathbb{E}[T_{\text{Top}}]$ yields the mean time to failure:

$$\text{MTTF} = \mathbb{E}[\text{time of the first system failure}] = \mathbb{E}[T_{\text{Top}}]$$

Using the definition of expected value, we obtain

$$\text{MTTF} = \mathbb{E}[T_{\text{Top}}] = \int_0^\infty t f_{\text{Top}}(t)\mathrm{d}t = \int_0^\infty \text{Reliability}(t)\mathrm{d}t$$

The latter equation follows from partial integration, and only holds if $\text{MTTF} < \infty$, which is the case in realistic systems. It follows from partial integration, as shown in this proof:

$$\text{MTTF} = \int_0^\infty t f_{\text{Top}}(t)\mathrm{d}t$$

$$= -\int_0^\infty -\frac{\mathrm{d}(1 - F_{\text{Top}}(t))}{\mathrm{d}t} t\,\mathrm{d}t \qquad\qquad \text{integration by parts}$$

$$= \left[-t(1 - F_{\text{Top}}(t))\right]_0^\infty + \int_0^\infty (1 - F_{\text{Top}}(t))\mathrm{d}t$$

$$= \int_0^\infty (1 - F_{\text{Top}}(t))\mathrm{d}t$$

$$= \int_0^\infty \text{Reliability}(t)\mathrm{d}t$$

Example 25

For a basic event e with failure rate λ, i.e., $T_e \sim \exp(\lambda)$ we have:

$$\text{Unreliability}(t) = \mathbb{P}[T_e < t] = 1 - e^{-\lambda t}$$
$$\text{Reliability}(t) = \mathbb{P}[X_e(t) = 0] = \mathbb{P}[T_e > t] = e^{-\lambda t}$$
$$\text{MTTF} = \mathbb{E}[T_e] = \frac{1}{\lambda}$$

11.6 Calculation of Dependability Metrics for Systems Without Repair

11.6.1 Point Availability/Reliability

Perhaps surprisingly, time-dependent analysis for systems without repair is quite similar to the time-independent case. In particular, all computation methods from the previous chapter work when substituting $p_i(t)$ for p_i.

- For tree-shaped fault trees, we can use the bottom up method to compute the top probability. Here, one propagates $p_i(t)$ from the leaves to the top. This yields a square-free polynomial in each step, i.e., a polynomial where all powers are 0 or 1.
- For minimal cut sets, take probabilities $p_i(t)$ instead of p_i.
- In the BDD method, we decorate the edges by $p_i(t)$.

If one wishes to compute the point availability or reliability for one specific point in time t_0, then one can simply treat the probabilities $p_1(t_0), p_2(t_0), \ldots p_n(t_0)$ as constants, and use the algorithms from the previous chapters. When treating the time t as a parameter, we can also obtain a closed-form solution for the top probability as a function of t. We illustrate how this works, both using the bottom up method and via a BDD.

Example 26 (Point availability with exponential failure distributions.)

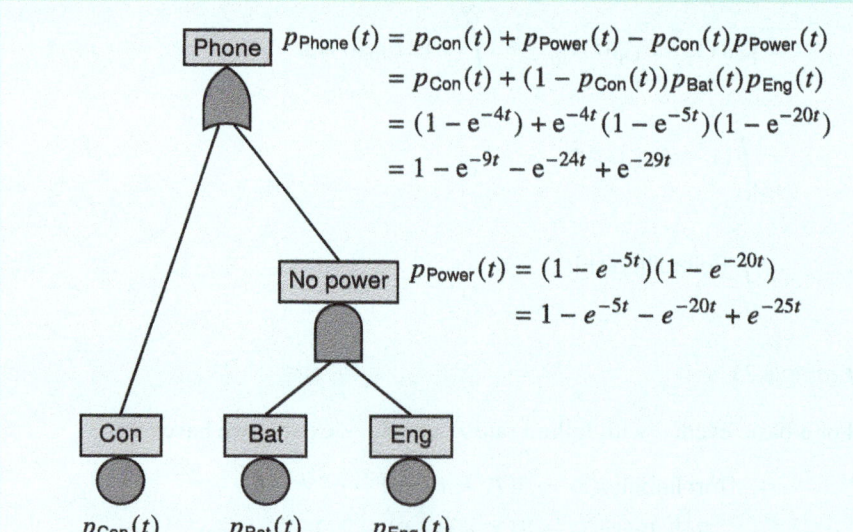

$$p_{\mathrm{Phone}}(t) = p_{\mathrm{Con}}(t) + p_{\mathrm{Power}}(t) - p_{\mathrm{Con}}(t)p_{\mathrm{Power}}(t)$$
$$= p_{\mathrm{Con}}(t) + (1 - p_{\mathrm{Con}}(t))p_{\mathrm{Bat}}(t)p_{\mathrm{Eng}}(t)$$
$$= (1 - e^{-4t}) + e^{-4t}(1 - e^{-5t})(1 - e^{-20t})$$
$$= 1 - e^{-9t} - e^{-24t} + e^{-29t}$$

$$p_{\mathrm{Power}}(t) = (1 - e^{-5t})(1 - e^{-20t})$$
$$= 1 - e^{-5t} - e^{-20t} + e^{-25t}$$

$$p_{\mathrm{Con}}(t) = 1 - e^{-4t} \qquad p_{\mathrm{Bat}}(t) = 1 - e^{-5t} \qquad p_{\mathrm{Eng}}(t) = 1 - e^{-20t}$$

This example shows the computation of the point availability for a tree-shaped fault tree using the bottom up algorithm in Sect. 10.3.

Example 27 (BDD for the road trip fault tree with exponential failure rates)

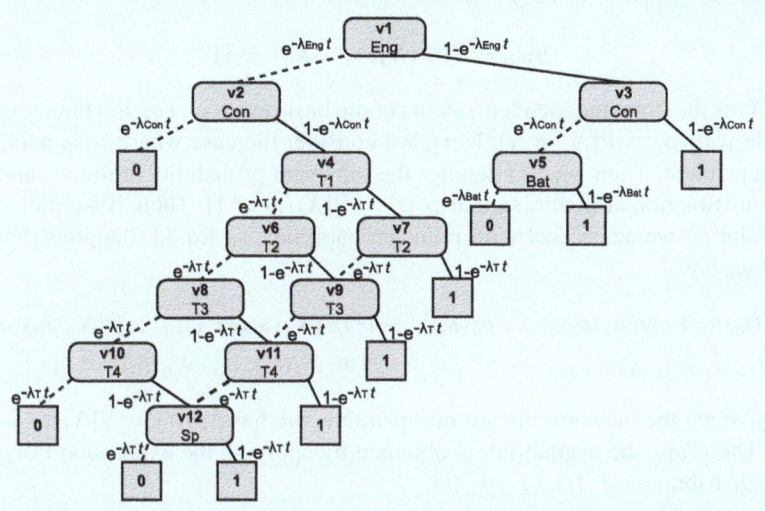

The BDD above shows the BDD for the road trip example, now decorated with exponential failure distributions. As before, the top probability can be computed by adding the probabilities of all paths, leading to the 1-leaf. The procedure uses similar equations as before, given in Sect. 10.5.1. Again, we simply substitute $p_i(t) = 1 - e^{-\lambda_i \cdot t}$ for p_i.

$$p^{\text{BDD}}(v_{12})(t) = (1 - e^{-\lambda_T \cdot t}) \cdot 1$$
$$p^{\text{BDD}}(v_{11})(t) = e^{\lambda_T \cdot t} p^{\text{BDD}}(v_{12}) + (1 - e^{\lambda_T \cdot t}) \cdot 1$$

$$....$$

$$p^{\text{BDD}}(v_1) = e^{\lambda_{\text{Eng}} \cdot t} p^{\text{BDD}}(v_2) + (1 - e^{\lambda_{\text{Eng}} \cdot t}) \cdot p^{\text{BDD}}(v_3)$$

General case. Chap. 10 provides several methods to compute the top probability in a fault tree. A careful analysis of these algorithms [3] shows that these algorithms can be executed symbolically, i.e., using the failure probabilities p_i as variables rather than numbers. It can also be shown that the result of this symbolic computation yields a function Poly over the basic event probabilities $p_1, \ldots, p_n$ to describe the probability of the top event, that is,

$$p_{\text{Top}} = \text{Poly}(p_1, \ldots, p_n)$$

The function Poly is a sum of products of probabilities and their complements, a.k.a. a square-free polynomial. Eq. 10.10 in Sect. 10.6 displays this function for the top probability in the road trip example. Key is that this function also works for time-dependent failure probabilities, as illustrated in Example 26. Recall from

Sect. 10.7 that p_{Top} is defined by applying the structure function Φ to random variables $X_1, \ldots, X_n$

$$p_{\text{Top}} = \mathbb{P}[\Phi(X_1, \ldots, X_n) = 1] \tag{11.3}$$

Here, X_i is the (time independent) status of the basic event e_i, i.e., is a binary random variable with $p_i = \mathbb{P}[X_i = 1]$. Next, we consider the case where p_{Top} and X_i are time-dependent. Then $p_{\text{Top}}(t)$ denotes the top event probability at time t, and X_i is the status function at time t, so that $p_i(t) = \mathbb{P}[X_i(t) = 1]$. Then, if we pick a fixed time point t_0, we again deal with point probabilities, so Eq. 11.3 applies. For each $t_0 \geq 0$ we have

$$\begin{aligned} p_{\text{Top}}(t_0) = \text{Poly}\big(p_1(t_0), \ldots, p_n(t_0)\big) &= \Phi^{\text{P}}(\mathbb{P}(X_1(t_0) = 1), \ldots, \mathbb{P}(X_n(t_0) = 1)) \\ &= \mathbb{P}[\Phi(X_1(t_0), \ldots, X_n(t_0)) = 1] \end{aligned}$$

Finally, since the basic events are unrepairable, we have $p_i(t) = \mathbb{P}[X_i(t) = 1] = F_i(t)$. Therefore, the availability is obtained by applying the expression Poly to the failure distributions $F_1(t), \ldots, F_n(t)$.

$$\begin{aligned} \text{Availability}^{\text{pt}}(t) &= 1 - \mathbb{P}[X_{\text{Top}}(t) = 1] \\ &= 1 - \mathbb{P}[\Phi(X_1(t), \ldots, X_n(t)) = 1] \\ &= 1 - \text{Poly}(p_1(t), \ldots, p_n(t)) \\ &= 1 - \text{Poly}(F_1(t), \ldots, F_n(t)) \end{aligned}$$

11.6.2   Computing the MTTF

Recall that the MTTF is given by

$$\text{MTTF} = \int\limits_0^\infty \text{Reliability}(t)dt$$

Computing this integral can be complex. For exponential distributions, one can find a primitive of the function Reliability(t), as illustrated in Example 28. Finding the primitive of other distributions can be more difficult.

Example 28 (MTTF of part of the road trip example)

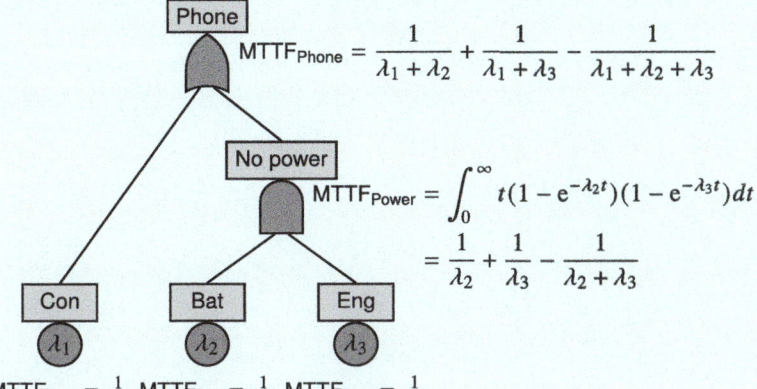

$$\text{MTTF}_{\text{Phone}} = \frac{1}{\lambda_1 + \lambda_2} + \frac{1}{\lambda_1 + \lambda_3} - \frac{1}{\lambda_1 + \lambda_2 + \lambda_3}$$

$$\text{MTTF}_{\text{Power}} = \int_0^\infty t(1 - e^{-\lambda_2 t})(1 - e^{-\lambda_3 t})dt$$

$$= \frac{1}{\lambda_2} + \frac{1}{\lambda_3} - \frac{1}{\lambda_2 + \lambda_3}$$

$$\text{MTTF}_{\text{Con}} = \frac{1}{\lambda_1} \quad \text{MTTF}_{\text{Bat}} = \frac{1}{\lambda_2} \quad \text{MTTF}_{\text{Eng}} = \frac{1}{\lambda_3}$$

The MTTF for **Phone** can be calculated as:

$$\text{MTTF(Phone)} = \int_0^\infty t\left(\left(1 - e^{-\lambda_1 t}\right) + \left(e^{-\lambda_1 t}\left(1 - e^{-\lambda_2 t}\right)\left(1 - e^{-\lambda_3 t}\right)\right)\right)dt$$

$$= \int_0^\infty t\left(1 - e^{-(\lambda_1 + \lambda_2)t} - e^{-(\lambda_1 + \lambda_3)t} + e^{-(\lambda_1 + \lambda_2 + \lambda_3)t}\right)dt$$

$$= \frac{1}{\lambda_1 + \lambda_2} + \frac{1}{\lambda_1 + \lambda_3} - \frac{1}{\lambda_1 + \lambda_2 + \lambda_3}$$

References

1. Ebeling CE (2009) An introduction to reliability and maintainability engineering, 2nd edn. Waveland Press
2. Rausand M, Barros A, Hoylan A (2020b) System reliability theory. Models, statistical methods, and applications. Wiley Series in Probability and Statistics, John Wiley & Sons
3. Lopuhaä-Zwakenberg M (2024) Fault tree reliability analysis via squarefree polynomials. In: Proceedings of the 12th international conference on model-based software and systems engineering, MODELSWARD, SCITEPRESS, pp 39–49
4. Haslwanter T (2022) Distributions of One Variable. Springer International Publishing, Cham, pp 105–138. https://doi.org/10.1007/978-3-030-97371-1

Dependability Metrics for Fault Trees with Repair

12

This chapter considers systems that can be repaired. We consider two cases: First, we study systems with *observable failures* that are repaired as soon as a failure occurs. A repair time distribution R_e describes the time it takes to repair the component. Second, we discuss systems with *unobservable failures*, also called hidden or unrevealed failures. These components are periodically tested, and, if the component has failed, it is repaired immediately.

In both cases, it is assumed that a repair brings a system to its pristine condition, without consequences for its future failure behavior. Moreover, we assume that each basic event has a dedicated repair unit, which repairs the component as soon as a failure is discovered.

This chapter covers the following cases.

- The point availability can be computed with methods that are similar to the point probabilistic model from Chap. 10. However, these methods only give readable expressions if the repair and failure distributions are both exponential.
- For systems with repair, it is common to study not only the point availability, but also the long-run average availability. We show that this metric can also be computed with methods that are similar to the point probabilistic model from Chap. 10.
- The reliability for repairable fault trees is computed by analyzing the stochastic process underlying the fault tree.
- Alternatively, Monte Carlo simulation can be used. Here, one takes a random sample for each basic event and then estimates the metric of interest for the fault tree.

© The Author(s), under exclusive license to Springer Nature Switzerland AG 2026   145
M. Stoelinga et al., *Concise Guide to Fault Tree Analysis*, Computer Science Foundations and Applied Logic, https://doi.org/10.1007/978-3-031-78287-9_12

Fig. 12.1 Model of a basic event with failure rate λ and repair rate μ

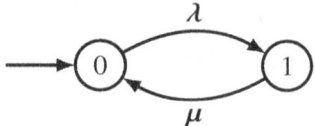

12.1 Examples: Exponential Failure and Repair Times

To study the influence of repairs, each basic event e is equipped with a repair time distribution R_e, in addition to its failure time distribution F_e. Thus, $R_e(t)$ is the probability that, after a failure, the basic event e is repaired within time t. Repair times are modeled by an exponential distribution whose parameter μ_e is called the *repair rate*. Then, $R_e(t) = 1 - e^{-\mu_e t}$. This section studies the case where all failure and repair times are exponential.

12.1.1 Basic Events

A component whose failure and repair times are both governed by exponential distributions can be described by a continuous-time Markov chain (CTMC) with two states. In state 0 of the CTMC, the system is operational, and in state 1, it has failed. The transition from 0 to 1 is governed by an exponential distribution with parameter λ, and the transition from 1 to 0 is governed by an exponential distribution with parameter μ. This is illustrated in Fig. 12.1.

Let $P_0(t)$ be the probability to be in the operational state 0 at time t and $P_1(t)$ the probability to be in the failed state 1 at time t. Markovian analysis shows that

$$P_0(t) = \frac{\mu}{\lambda + \mu} + \frac{\lambda}{\lambda + \mu} e^{-(\lambda+\mu)t}$$

$$P_1(t) = \frac{\lambda}{\lambda + \mu} - \frac{\lambda}{\lambda + \mu} e^{-(\lambda+\mu)t}$$

Availability. Note that $P_1(t)$ is the point unavailability at time t. Thus we have

$$\text{Unavailability}_e^{\text{pt}}(t) = \frac{\lambda}{\lambda + \mu} - \frac{\lambda}{\lambda + \mu} e^{-(\lambda+\mu)t}$$

As we will show, the long-run average availability is obtained[1] by taking the limit. Note that

$$\text{Availability}_e^{\text{lr}} = \frac{\text{MTTF}_e}{\text{MTTF}_e + \text{MTTR}_e}$$

---

[1] This definition applies since the limit in question exists.

This can be seen as follows. Recall that the mean time to failure is given by $\text{MTTF}_e = \mathbb{E}[T_e] = 1/\lambda$. Similarly, the mean time to repair equals $\text{MTTR}_e = \mathbb{E}[R_e] = 1/\mu$.

$$\frac{\text{MTTF}_e}{\text{MTTF}_e + \text{MTTR}_e} = \frac{1/\lambda}{1/\lambda + 1/\mu} = \frac{\mu}{\lambda + \mu} = \text{Availability}_e^{\text{lr}}$$

Note that the CTMC above is useful to derive the equations for $P_0(t)$ and $P_1(t)$, but it does not play a role in the availability computation.

> **Example 29**
>
> For a component with failure rate $\lambda = 1/3$ and repair rate $\mu = 1$ we have
>
> $$\text{Availability}^{\text{lr}} = \frac{\text{MTTF}}{\text{MTTF} + \text{MTTR}} = \frac{3}{3+1} = \frac{3}{4}$$
>
> Thus, the component is available 75% of the time.

Reliability. Recall that the reliability of a system is the probability that the system does not fail before a given time point t. For basic events, the reliability can be computed by ignoring repairs.

This is because whether a basic event e has failed before time t is independent of any subsequent repairs. However, this does not hold at the system level, as repairing the component could have entirely averted a system failure.

12.1.2 Intermediate and Top Events

Availability. Just like for the irreparable case, analyzing the availability for repairable fault trees closely mirrors the approach for the time-independent case. This holds both for the point and the long-run average availability. Therefore, the calculation methods for the point probabilistic model can be exploited for repairable fault trees as well: for tree-shaped fault trees, we can use the bottom-up method, for other cases the minimal cut set approximations, and BDD computations. We illustrate the point and long-run average availability on an AND-gate.

Consider an AND-gate E with two children e_1 and e_2, where failure and repair rates of e_i are λ_i and μ_i, respectively. Since the point availabilities are probabilistically independent, they can be multiplied to obtain the point unavailability of the AND-gate.

$$
\begin{aligned}
&\text{Unavailability}_{\text{AND}(e_1,e_2)}^{\text{pt}}(t) \\
={}& \text{Unavailability}_{e_1}^{\text{pt}}(t) \cdot \text{Unavailability}_{e_2}^{\text{pt}}(t) \\
={}& \left(\frac{\lambda_1}{\lambda_1 + \mu_1} - \frac{\lambda_1}{\lambda_1 + \mu_1} e^{-1(\lambda_1+\mu_1)t} \right) \left(\frac{\lambda_2}{\lambda_2 + \mu_2} - \frac{\lambda_2}{\lambda_2 + \mu_2} e^{-1(\lambda_2+\mu_2)t} \right)
\end{aligned}
$$

Note this equation has the form $p_E(t) = p_1(t)p_2(t)$, showing that the equation is an immediate generalization of the rules for gate probabilities in Table 10.1. As before, these rules only hold if all children of an event are modules. The rules for the OR- and VOTING-gates can be generalized in a similar way.

Moreover, since the point unavailability of the AND-gate is the product of the point availabilities of its children, the same holds for its long-run average availability: this is also the product of the long-run average availability of the children. The reason is that the limit distributes over multiplication:

$$
\begin{aligned}
\text{Unavailability}^{\text{lr}}_{\text{AND}(e_1,e_2)} &= \lim_{t\to\infty} \text{Unavailability}^{\text{pt}}_{\text{AND}(e_1,e_2)}(t) \\
&= \lim_{t\to\infty} \text{Unavailability}^{\text{pt}}_{e_1}(t) \cdot \text{Unavailability}^{\text{pt}}_{e_2}(t) \\
&= \lim_{t\to\infty} \text{Unavailability}^{\text{pt}}_{e_1}(t) \cdot \lim_{t\to\infty} \text{Unavailability}^{\text{pt}}_{e_2}(t) \\
&= \text{Unavailability}^{\text{lr}}_{e_1} \cdot \text{Unavailability}^{\text{lr}}_{e_2} \\
&= \frac{\lambda_1}{\lambda_1 + \mu_1} \frac{\lambda_2}{\lambda_2 + \mu_2}
\end{aligned}
$$

This equation is also an immediate generalization of the rules for gate probabilities in Table 10.1. Again, similar rules for the OR-and VOTING-gates can be derived. We can apply these rules to obtain the point and long-run average availability in the road trip example.

Example 30 (Road trip example with exponential failures and repairs.)

Section 10.6 showed that the point unavailability at time t equals

$$
\begin{aligned}
p_{Top}(t) &= \text{Unavailability}^{\text{pt}}_{Top}(t) \\
&= \big(p_1(t) + p_2(t) - p_1(t)p_2(t)\big)p_3(t) + \\
&\quad p_1(t)\overline{p_3(t)}\left(1 - \overline{p_4(t)}^5 - 5p_4(t)\overline{p_4(t)}^4\right)
\end{aligned}
$$

where $p_i(t) = \frac{\lambda_i}{\lambda_i + \mu_i} - \frac{\lambda_i}{\lambda_i + \mu_i}e^{-(\lambda_i+\mu_i)t}$. Moreover, we have

$$
\begin{aligned}
\text{Unavailability}^{\text{lr}} &= \left(\frac{\lambda_1}{\lambda_1 + \mu_1} + \frac{\lambda_2}{\lambda_2 + \mu_2} - \frac{\lambda_1}{\lambda_1 + \mu_1}\cdot\frac{\lambda_2}{\lambda_2 + \mu_2}\right)\frac{\lambda_3}{\lambda_3 + \mu_3} \\
&\quad + \frac{\lambda_1}{\lambda_1 + \mu_1}\frac{\mu_3}{\lambda_3 + \mu_3}\left(1 - \left(\frac{\mu_4}{\lambda_4 + \mu_4}\right)^5 - 5\frac{\lambda_4}{\lambda_4 + \mu_4}\left(\frac{\mu_4}{\lambda_4 + \mu_4}\right)^4\right)
\end{aligned}
$$

Fig. 12.2 CTMC underlying
an AND-gate

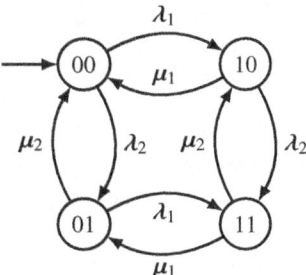

Reliability. Whereas the reliability and the availability coincide for systems without repairs, and the reliability for basic events can be obtained by ignoring repairs, the situation for intermediate events is more complex.

Consider an AND-gate E with children e_1 and e_2. We would like to calculate the reliability at time $t = 100$. However, it may happen that before this time, e_1 has failed and got repaired. If e_2 continued working, then the failure of e_1 was not visible at system level. Therefore, computing the reliability of E is performed on the underlying Markov process. If all failure and repair distributions are exponential, then the Markov process is a continuous-time Markov chain (CTMC).

The CTMC underlying $E = \mathsf{AND}(e_1, e_2)$ is displayed in Fig. 12.2. The CTMC is obtained by considering all status vectors of E, together with their failure behavior. For example, the arrow from state 00 to 10 indicates that basic event e_1 failed, i.e., moved from status 0 to status 1. Since this transition is governed by failure rate λ_1, that rate is written along the arrow. In state 10, e_1 can either be repaired (leading the fault tree back to its initial status vector 00; with rate μ_1), or e_2 can fail. The latter happens with rate e_2's failure rate λ_2 and leads the system to 11.

Since $E = \mathsf{AND}(e_1, e_2)$, the only failed state in the CTMC is 11. Therefore, the reliability of E is computed as the absorption time in state 11: what is the probability that the first visit to state 11 occurs before time t.

If we consider $E = \mathsf{OR}(e_1, e_2)$, then we can use the same underlying CTMC. However, the OR-gate gives rise to failure states 10, 01, 11. Thus, the reliability of the OR-gate can be calculated as the first time to visit one of the states 10, 01, or 11. Note that, to analyze the OR-gate, the state 11 and all its transitions can be removed from the CTMC. To visit state 11, one must visit either 10 or 01 before, making the transition to 11 superfluous.

Markov chains are a very well-established model, with numerous efficient techniques for their analysis [1]. The construction of the underlying CTMC critically relies on the memoryless property of the exponential failure and repair distributions, and does not generalize easily to other probability distributions.

12.2 Examples: Beyond Exponentials

12.2.1 Basic Events

For the general case, we assume that a component's failure is governed by a failure time distribution F_e and its repair time distribution is governed by a repair time distribution R_e. The component's failure behavior can be described by a semi-Markov process with two states.

In state 0, the system is operational, and in state 1, it has failed. The transition from 0 to 1 is governed by F_e. That is, the time to stay in state 0 is a random variable, namely the failure time T_e, whose distribution is F_e. This is illustrated in Fig. 12.3. Similarly, the transition from 1 to 0 is governed by R_e. The time to stay in state 1 is described by the random variable Rep, whose distribution is R_e.

Availability. As before, the point unavailability at time t is the probability to be in state 1 of the semi-Markov process at time t.

$$P_0(t) = \text{Availability}^{\text{pt}}(t)$$
$$P_1(t) = \text{Unavailability}^{\text{pt}}(t)$$

In general, it can be difficult to obtain closed-form expressions for $P_0(t)$ and $P_1(t)$, making the point availability difficult to compute. However, under mild assumptions, the semi-Markov process has a steady-state distribution that corresponds to the long-run average [1]. The steady-state distribution is obtained as the limit of P_0 and P_1 when time goes to infinity. Sufficient conditions for the limit to exist are that F_e and R_e are not both deterministic and that their expected values are defined and finite. We define

$$S_0 = \lim_{t \to \infty} P_0(t)$$
$$S_1 = \lim_{t \to \infty} P_1(t)$$

Note that $S_i \in \{0, 1\}$ represents the fraction of time that the process spends in state i, when time goes to infinity, i.e., the fraction of time that the system is operational in the long run. As such, S_0 is strongly related to the mean time to failure.

$$\text{Availability}^{\text{lr}} = S_0 = \frac{\mathbb{E}[T_e] + \mathbb{E}[\text{Rep}_e]}{\mathbb{E}[\text{Rep}_e]} = \frac{\text{MTTF}}{\text{MTTF} + \text{MTTR}}$$

$$\text{Unavailability}^{\text{lr}} = S_1 = \frac{\mathbb{E}[T_e] + \mathbb{E}[\text{Rep}_e]}{\mathbb{E}[F_e]} = \frac{\text{MTTR}}{\text{MTTF} + \text{MTTR}}$$

Fig. 12.3 Model of a basic event with failure distribution F_e and repair distribution R_e

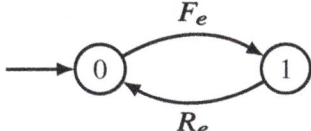

Example 31

For a component with failure rate $\lambda = 1/3$ and a deterministic repair time $D = 1$, we have

$$\text{Availability}^{lr} = \frac{\text{MTTF}}{\text{MTTF} + \text{MTTR}} = \frac{3}{3+1} = \frac{3}{4}$$

Thus, the component is available 75% of the time. This is the same as for the case with exponential repair. As shown above, it is the same for any repair distribution with a mean value of 1: In the long run, the exact failure distribution averages out. The only value that matters is its average, and the exact distribution around the average is irrelevant.

Reliability. As in the non-repairable case, the reliability of a component can be obtained by ignoring repairs. The reason is exactly the same: at event-level no repairs can happen before the event has failed.

12.2.2   Intermediate and Top Events

Availability. We can treat the point and long-run availabilities in the same way as in the case without repairs, which in turn is similar to the point probabilistic model.

The reason is that the gate rules in Table 10.1 only rely on the probabilistic independence of the probabilities, not on the shape of the distributions. For an AND-gate E with two basic events as children

$$p_E(t) = p_1(t) \cdot p_2(t)$$

The exact shape of the distribution is not relevant. Reformulated as point probabilities, we obtain:

$$\text{Unavailability}^{pt}_{\text{AND}(e_1,e_2)}(t) = \text{Unavailability}^{pt}_{e_1}(t) \cdot \text{Unavailability}^{pt}_{e_2}(t)$$

Note that the equation above is again an immediate generalization of the rules for gate probabilities in Table 10.1. The rules for OR-and voting gates can be generalized in a similar way.

For the long-run average, we take the limit of the point probabilities. This limit commutes with the linear function, which means that we can take the limit of the basic events and use their values in the probabilistic structure function to obtain the long-run average.

We show this for an AND-gate with children e_1 and e_2.

$$
\begin{aligned}
\text{Unavailability}^{\text{lr}}_{\text{AND}(e_1,e_2)} &= \lim_{t\to\infty} \text{Unavailability}^{\text{pt}}_{\text{AND}(e_1,e_2)}(t) \\
&= \lim_{t\to\infty} \text{Unavailability}^{\text{pt}}_{e_1}(t) \cdot \text{Unavailability}^{\text{pt}}_{e_2}(t) \\
&= \lim_{t\to\infty} \text{Unavailability}^{\text{pt}}_{e_1}(t) \cdot \lim_{t\to\infty} \text{Unavailability}^{\text{pt}}_{e_2}(t) \\
&= \text{Unavailability}^{\text{pt}}_{e_1} \cdot \text{Unavailability}^{\text{pt}}_{e_2} \\
&= \frac{\text{MTTF}_{e_1}}{\text{MTTF}_{e_1} + \text{MTTR}_{e_2}} \cdot \frac{\text{MTTF}_{e_2}}{\text{MTTF}_{e_1} + \text{MTTR}_{e_2}}
\end{aligned}
$$

We use the argumentation above to obtain the long-run average availability in the road trip example with exponential failures and deterministic repairs.

> ### Example 32 (Road trip example with exponential failures and deterministic repairs.)
>
> Suppose the component i has an exponential failure rate λ_i and a deterministic repair time D_i. Let d_i denote $1/D_i$. Recall that
>
> $$
> \begin{aligned}
> p_{Top}(t) = \text{Unavailability}^{\text{pt}}_{Top}(t) \\
> = \big(p_1(t) + p_2(t) - p_1(t)p_2(t)\big)p_3(t) + \\
> p_1(t)\overline{p_3(t)}\left(1 - \overline{p_4(t)}^5 - 5p_4(t)\overline{p_4(t)}^4\right)
> \end{aligned}
> $$
>
> By taking the limit for t to infinity we can see
>
> $$
> \begin{aligned}
> \text{Unavailability}^{\text{lr}} = &\left(\frac{\lambda_1}{\lambda_1 + d_1} + \frac{\lambda_2}{\lambda_2 + d_2} - \frac{\lambda_1}{\lambda_1 + d_1}\frac{\lambda_2}{\lambda_2 + d_2}\right)\frac{\lambda_3}{\lambda_3 + d_3} \\
> &+ \frac{\lambda_1}{\lambda_1 + d_1}\frac{d_3}{\lambda_3 + d_3}\left(1 - \left(\frac{d_4}{\lambda_4 + d_4}\right)^5 - 5\frac{\lambda_4}{\lambda_4 + d_4}\left(\frac{d_4}{\lambda_4 + d_4}\right)^4\right)
> \end{aligned}
> $$

Reliability. Section 12.1.2 illustrated that the reliability for fault trees with exponential failures and repairs can be computed via a CTMC, describing the probabilistic behavior of the entire fault tree. This CTMC construction relies critically on the memoryless properties of the exponential distribution.

In the non-exponential case, one must create a semi-Markov process for each basic event and consider the cumulative behavior of all these processes together. In general, the joint behavior of such semi-Markov processes is not easy [2].

Fig. 12.4 Model of a basic
event with failure
distribution F_e and repair
distribution R_e

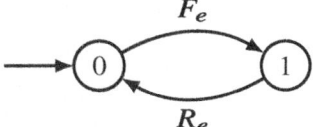

12.3 The Time-Dependent Probabilistic Model with Repair

This section provides the probabilistic model needed to define the dependability metrics for systems with repairs. We define the status function, as well as the failure and repair times as random variables. The dependability metrics are then defined in terms of these random variables.

12.3.1 Status Function for Repairable Systems

The starting point for fault tree analysis of repairable systems is again the *status function* $X_E(t)$ that describes the status of event E at time t:

$$X_E(t) = \begin{cases} 1 & \text{Event } E \text{ is failed at time } t \\ 0 & \text{Event } E \text{ is operational at time } t \end{cases}$$

In non-repairable systems, X_E transitions to 1 only once. In contrast, for repairable systems, X_E can switch between 0 and 1 multiple times, see Fig. 12.5. Recall that X_E is a random variable, so $\mathbb{P}[X_E(t) = 1]$ is the probability that event E is failed at time t, and $\mathbb{P}[X_E(t) = 0]$ is the probability that event E is operational at time t.

As before, the status function X_E for an intermediate event E and top event *Top* can be derived from the structure function

$$X_E = \Phi_F(X_1, \dots, X_n, E)$$
$$X_{Top} = \Phi_F(X_1, \dots, X_n)$$
$$\mathbb{P}[X_E(t) = 1] = \mathbb{P}[\Phi_F(X_1(t), \dots, X_n(t), E) = 1]$$
$$\mathbb{P}[X_{Top}(t) = 1] = \mathbb{P}[\Phi_F(X_1(t), \dots, X_n(t)) = 1]$$

12.3.2 Failure and Repair Time Distributions

For repairable systems, we derive both the failure time and the repair time distributions. The failure time for repairable systems is defined as for the irreparable case: Given X_E, the *failure time* T_E is the point in time when E fails for the first time, see Fig. 12.5.

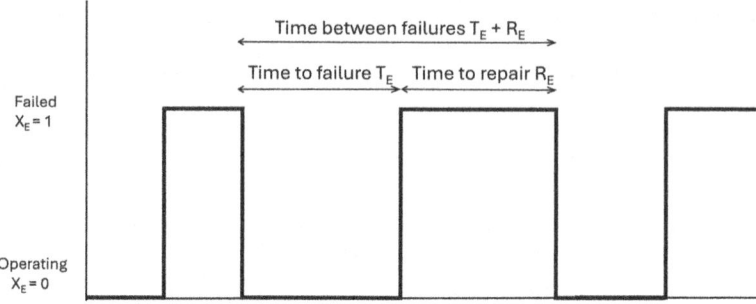

Fig. 12.5 Repairable systems: status variable X_E, time to failure T_E, time to repair R_E, time between failures $T_E + R_E$

$$T_E = \min\{t \in [0, \infty) \mid X_E(t) = 1\}$$
$$F_E(t) = \mathbb{P}[\text{Event } E \text{ fails before time } t] = \mathbb{P}[T_E < t]$$

Commonly, F_E is called the *failure (time) distribution* of E. The repair time is the point in time where the system was first repaired after the last failure.

$$\text{Rep}_E(t) = \min\{t - T_E \in [0, \infty) \mid X_E(t) = 0, t > T_E\}$$

We write R_E for the distribution function of Rep, yielding the probability that E was repaired before time t, see Fig. 12.5. We define

$$R_E(t) = \mathbb{P}[\text{Event } E \text{ is repaired before time } t + T_E] = \mathbb{P}[\text{Rep}_E < t]$$

Commonly, R_E is called the *repair (time) distribution* of E. Again, we equip the basic events with repair time distributions and we derive repair times for other events. We also consider the probability density function of R_E, indicating how fast R_E changes over time.

$$r_E(t) = R'_E(t) \text{ and } R_E(t) = \int_0^t r_E(x)\mathrm{d}x$$

The above definitions apply to the basic, intermediate and top events. As before, we assign failure time distributions to all basic events e and calculate failure time distributions and other dependability metrics for other events, especially the top event.

12.4   Dependability Metrics for Systems with Repair

Availability. For repairable systems, several notions of availability are relevant. Recall that *point availability* is the probability that the system is operational at a given time point t.

$$\text{Availability}^{\text{pt}}(t) = \mathbb{P}[\text{system is operational at time } t]$$
$$= \mathbb{P}[X_{Top}(t) = 0]$$
$$= \mathbb{P}[\Phi(X_1(t), \dots, X_n(t)) = 0]$$

The *average availability*, also called *interval* or *mission* availability is the percentage of time that a system is operational during a time interval $[0, T]$.

$$\text{Availability}^{\text{int}}(T) = \frac{1}{T} \int_0^T \text{Availability}^{\text{pt}}(t) dt$$

Finally, the *long-run availability* or *steady-state availability*, often abbreviated as *availability*, or *uptime* is the average percentage of time that a system is operational.

$$\text{Availability}^{\text{lr}} = \lim_{T \to \infty} \text{Availability}^{\text{int}}(T)$$

A well-known equation states that the long-run availability equals the average fraction of time that the system is up:

$$\text{Availability}^{\text{lr}} = \frac{\text{MTTF}}{\text{MTTF} + \text{MTTR}}$$

Remark 2 The long-run availability is defined as the limit of the interval availability, not as the limit of the point availability. The reason is that the limit for the point availability may not exist in cases where the limit for the interval availability does exist. For example, if both the failure and repair times are deterministic (say they are both equal to 1), then the point availability does not have a limit, whereas the interval availability has a limit $1/2$. Another typical example is the tested repair model from Chap. 10. However,

$$\text{if } \lim_{T \to \infty} \text{Availability}^{\text{pt}}(T) \text{ exists,}$$

$$\text{then } \lim_{T \to \infty} \text{Availability}^{\text{pt}}(T) = \lim_{T \to \infty} \text{Availability}^{\text{int}}(T)$$

Reliability. The definition of reliability is the same as for irreparable systems. One considers the first point in time where the top event failed.

$$\text{Reliability}(t) = \mathbb{P}[\text{system does not fail before time } t] = 1 - F_{Top}(t) = \mathbb{P}[T_{Top} > t]$$
$$\text{Unreliability}(t) = \mathbb{P}[\text{fails before time } t] = 1 - \text{Reliability}(t) = F_{Top}(t)$$

As before, the reliability is also called the *survivor function*, asking whether the system has survived the first t time units. Clearly, in reparable systems, components may have failed and repaired before a top failure occurred.

MTTF. Recall that the MTTF is defined by

$$\text{MTTF} = \mathbb{E}[\text{first system failure appears at time } T]$$
$$= \mathbb{E}[T_{Top}]$$
$$= \int_0^\infty t f_{Top}(t) dt$$
$$= \int_0^\infty \text{Reliability}(t) dt$$

MTTR and MTBF. Mean time to repair (MTTR) is the average time to troubleshoot and remedy an issue. That is, the MTTR is the expected value of the repair time distribution. The mean time before failure (MTBF) is the sum of the MTTF and the MTTR.

$$\text{MTTR} = \mathbb{E}[R_{Top}]$$
$$\text{MTBF} = \text{MTTF} + \text{MTTR} = \mathbb{E}[F_{Top}] + \mathbb{E}[R_{Top}]$$

12.5 Calculation of Dependability Metrics for Systems with Repair

12.5.1 Point Availability

The computation of point availability for repairable systems is also similar to the case without repairs (and hence to the point probabilistic case), except that we need to use different distributions at the basic event level.

Chapter 10 provides several methods to compute the top probability in a fault tree. Recall that the point availability function Poly over the BE probabilities $p_1, \ldots, p_n$ describes the probability of the top event, that is,

$$p_{Top} = \text{Poly}(p_1, \ldots, p_n)$$

Example 33 (Long-run availability with exponential failures and repairs.)

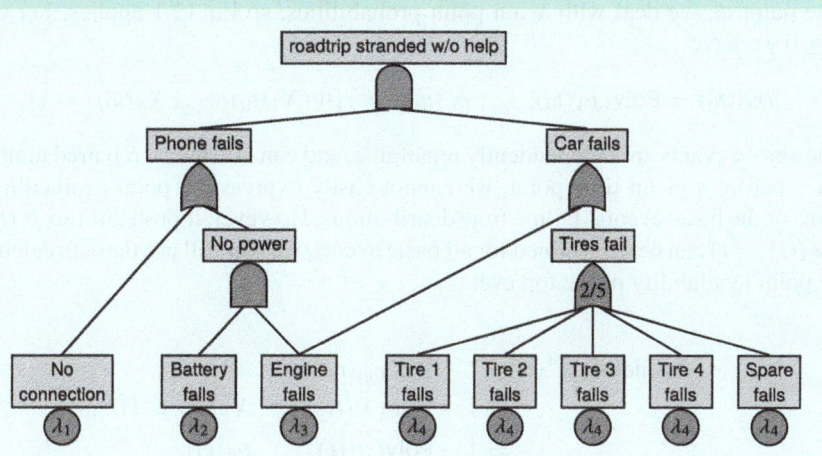

Assume that all basic events are governed by exponential failure distributions with the parameters λ_i, and exponential repair distributions with parameters μ_i.

$$p_{Top} = (p_1 + p_2 - p_1 p_2)p_3 + p_1\overline{p_3}\left(1 - \overline{p_4}^5 - 5p_4\overline{p_4}^4\right). \qquad (12.1)$$

To compute the long-run average availability, we substitute

$$p_i = \frac{\lambda_i}{\lambda_i + \mu_i}$$

Note that, in fact, $p_i(t) = F_{e_i}(t)$. Since $p_i(t)$ is exponentially distributed with parameter λ_i we have $p_i(t) = F_{e_i}(t) = 1 - e^{-\lambda t}$. Substitution yields

$$p_{Top}(t) = F_{Top}(t) =$$
$$\left(1 - e^{-\lambda_1 t} + 1 - e^{-\lambda_2 t} - (1 - e^{-\lambda_1 t})(1 - e^{-\lambda_2 t})\right)(1 - e^{-\lambda_3 t}) +$$
$$(1 - e^{-\lambda_1 t})e^{-\lambda_3 t}\left(1 - (e^{-\lambda_4 t})^5 + 5(1 - e^{-\lambda_4 t})(e^{-\lambda_4 t})^4\right)$$

Equation 10.10 in Example 26 displays this polynomial for the top probability in the road trip example. Key is that this polynomial also works for time-dependent failure probabilities, as illustrated in Example 26. Recall from Sect. 10.7 that p_{Top} is defined by applying the structure function Φ to random variables $X_1, \ldots, X_n$

$$p_{Top} = \mathbb{P}[\Phi(X_1, \ldots, X_n) = 1] \qquad (12.2)$$

Here, X_i is the (time-independent) status of the basic event e_i, i.e., is a binary random variable with $p_i = \mathbb{P}[X_i = 1]$. Next, we consider the case where p_{Top} and X_i be time-dependent. Then $p_{Top}(t)$ denotes the top event probability at time t, and X_i is the status function at time t, so that $p_i(t) = \mathbb{P}[X_i(t) = 1]$. Then, if we pick a fixed time point t_0, we deal with again point probabilities, so Eq. 12.1 applies. For each $t_0 \geq 0$ we have

$$p_{Top}(t_0) = \mathsf{Poly}(p_1(t_0), \ldots, p_n(t_0)) = \mathbb{P}[\Phi(X_1(t_0), \ldots, X_n(t_0)) = 1]$$

Since basic events are independently repairable, and can fail and be repaired multiple times before a given time point, we cannot easily express the point availability in terms of the basic events' failure time distributions. However, if probabilities $p_i(t) = \mathbb{P}[X_i(t) = 1]$ can be determined for all basic events, we can still use these to calculate the point availability of the top event:

$$\begin{aligned}
\mathsf{Availability}^{\mathsf{pt}}(t) &= 1 - \mathbb{P}[X_{Top}(t) = 1] \\
&= 1 - \mathbb{P}[\Phi(X_1(t), \ldots, X_n(t)) = 1] \\
&= 1 - \mathsf{Poly}(p_1(t), \ldots, p_n(t))
\end{aligned}$$

12.6   Repairable System with Unobservable Failures

Components that are amenable to unobservable failures need to be tested periodically. If a failure is found, then the component is repaired.

We consider a basic event with an exponential failure rate λ and a test interval θ. That is, the component is tested every θ time units. Note that this approach corresponds to the periodically tested model from Sect. 10.1.

Point unavailability. The point unavailability is given by

$$\mathsf{Unavailability}^{\mathsf{pt}}(t) = 1 - e^{\lambda(t - \theta \lfloor t/\theta \rfloor)}$$

Here $\lfloor x \rfloor$ rounds the number x down to the nearest integer. The point availability can be analyzed with the method in Sect. 12.5.1.

Long-Run Average Availability. For a component with failure rate λ and testing time θ, the average availability is given by

$$\begin{aligned}
\mathsf{Availability}^{\mathsf{int}}(\theta) &= \frac{1}{\theta} \int_0^\theta (1 - e^{-\lambda t}) dt \\
&= \frac{1}{\theta} [t + \frac{e^{-\lambda t}}{\lambda}]_0^\theta \\
&= 1 - \frac{1 - e^{\lambda \theta}}{\lambda \theta}
\end{aligned}$$

This availability gives us the long-run availability by noting that each inspection interval has the same average interval availability, giving:

$$\text{Availability}^{\text{lr}} = \lim_{T \to \infty} \text{Availability}^{\text{int}}(T)$$

$$= \lim_{n \to \infty} \text{Availability}^{\text{int}}(n\theta)$$

$$= \lim_{n \to \infty} \text{Availability}^{\text{int}}(\theta)$$

$$= 1 - \frac{1 - e^{\lambda\theta}}{\lambda\theta}$$

It should be noted that, if multiple components use the same inspection interval, they are not stochastically independent, if the components start at the same time.

12.6.1 Reliability

Computation of the reliability for repairable systems is complex. It can happen that several components fail and get repaired before a system failure occurs. Many computation methods rely on the underlying semi-Markov process of the fault tree [2].

In the case where all failure and repair times are exponential, the semi-Markov process is a continuous-time Markov chain, which can be analyzed with Markov solvers. Section 12.1.2 showed that the reliability for an AND-gate and an OR-gate with basic events as children is calculated as the absorption time in CTMC. These structures do actually propagate over gates in a module, where the CTMC of an intermediate event is obtained as the Cartesian product of the CTMCs of its children.

For the non-exponential case, the underlying stochastic process is a semi-Markov process, which is more complex. In such cases, one can resort to Monte Carlo simulation.

12.7 Monte Carlo Simulation

When the failure or repair distributions of components are too complex for algebraic analysis, it may be useful to use Monte Carlo simulation to estimate the reliability metrics.

Monte Carlo simulation is a very general concept: It can be used to analyze practically any metric that can be measured, and a wide variety of statistical procedures can be used to estimate how precise the calculated result is.

For illustration, we present a simple algorithm (formalized in Algorithm 1) for estimating the reliability of a fault tree F up to a time horizon T: Intuitively, we simulate the fault tree a given number of times N. Each simulation is stopped when either F has experienced a simulated failure, or we reach the time horizon. The fraction of simulations that reach a failure provides a point estimate for the unreliability of F at time T.

Each simulated run begins with all basic events operational (no failures). We sample a simulated failure time for each event e, following its failure time distribution F_e. We now increment the simulated time to the failure of whichever event fails first, mark its failure, and sample a repair time following the distribution R_e. We then repeatedly proceed, each time to the occurrence of the next simulated failure or repair, keeping track of which basic events are currently failed. When the failed basic events cause the fault tree to fail, or we exceed the time horizon, the simulated run ends and we mark whether a failure of F occurred before the time expired.

Algorithm 1 Simple Monte Carlo simulation for Reliability

Input: Fault tree F, Time horizon T, Simulation count N
Output: Estimated Reliability $R_F(T)$
procedure SimReliability(F, t, N)
  FailCount $\leftarrow 0$
  for $i \leftarrow 1, \ldots, N$ **do**
    FailedBEs $\leftarrow \emptyset$
    Events $\leftarrow \emptyset$
    for all $e \in BE(F)$ **do**
      $r \leftarrow$ Rand(F_e)
      Events $\leftarrow$ Events $\cup \{\langle r, \text{Fail}, e \rangle\}$
    end for
    $t \leftarrow 0$
    while $t < T$ & $\Phi_F(\text{FailedBEs}) = 0$ **do**
      $e \leftarrow \langle r, v, e \rangle \in$ Events **where** $\forall \langle r', v', e' \rangle \in$ Events $: r' \geq r$
      Events $\leftarrow$ Events $\setminus \{e\}$
      $t \leftarrow r$
      if $v = $ Fail **then**
        FailedBEs $\leftarrow$ FailedBEs $\cup \{e\}$
        $r \leftarrow$ Rand(R_e)
        Events $\leftarrow$ Events $\cup \{\langle t + r, \text{Repair}, e \rangle\}$
      else
        FailedBEs $\leftarrow$ FailedBEs $\setminus \{e\}$
        $r \leftarrow$ Rand(F_e)
        Events $\leftarrow$ Events $\cup \{\langle t + r, \text{Fail}, e \rangle\}$
      end if
    end while
    if $t \leq T$ **then**
      FailCount $\leftarrow$ FailCount $+ 1$
    end if
  end for
  $R_F(T) = 1 - \text{FailCount}/N$
end procedure

Variations on Monte Carlo Simulation. The algorithm described above is a very simple case: It measures only the reliability, and provides a point estimate with no confidence information. In practice, more complex variations are generally used:

Metrics: Any metric can be estimated as long as it can be measured in the simulation runs. For reliability, as described, we measure the number of simulations that end in failure. To estimate MTTF, we run each simulation until a failure occurs and measure its time. For unavailability, we measure the fraction of time within a simulation run that the fault tree is in a failed state. More complex measures can also be estimated. For example, one can track which events are failed when a simulation run shows a Top failure, to estimate importance metrics of the different events.

Statistical Metrics: The simple algorithm provides only a point estimate, with no information about how reliable this metric is. Practical systems generally provide more informative estimates, such as confidence intervals.

Stopping Criterion: The simple algorithm runs a fixed number of simulations. This is not very user-friendly, as it is hard to estimate in advance how many simulations are needed. Practical approaches generally provide more options, such as running simulations until the metric is estimated (with high confidence) to a given precision, or until the simulation shows with confidence whether the system meets or fails some dependability requirement.

Failure and Repair Distributions: The failure and repair times can, in principle, follow arbitrary probability distributions. It is even possible to use distributions that vary over time (e.g., for components that experience wear and tear) or system state (e.g., when failures of some components affect the failure rates of others). Different simulation algorithms offer varying amounts of flexibility.

Pitfalls of Monte Carlo Simulation. While Monte Carlo simulation is practical and flexible analysis technique, there are some potential pitfalls to consider:

Computational Requirements: Monte Carlo simulation can require large amounts of time to reach an estimate with high confidence, particularly when the system being analyzed is highly reliable and/or a high degree of precision is required: Consider a system with a failure probability of one in a million. This means that only one simulated failure is observed in approximately every million simulations. Refining this into a precise estimate of the reliability can require billions of simulations, which may take a long time. Special techniques for rare event simulation can reduce the simulations required [3,4], but these only work for restricted probability distributions and/or fault tree structures.

Probabilistic Results: Monte Carlo simulation can only estimate the measure of interest. It does not provide an exact value, nor a guarantee of over- or under-estimating the true metric. This may not be acceptable for some regulations.

Repeatability: Because Monte Carlo simulation works by drawing random failure and repair times, repeating the same estimation can give a different result. This can complicate questions such as comparing different design options, where statistical methods like hypothesis testing must be used to distinguish between actual changes in the metric and mere statistical artifacts.

References

1. Puterman ML (2014) Markov decision processes: discrete stochastic dynamic programming. John Wiley & Sons. https://doi.org/10.1002/9780470316887
2. Monti RE, Budde CE, D'Argenio PR (2020) A compositional semantics for repairable fault trees with general distributions. In: Albert E, Kovács L (eds) LPAR 2020: 23rd international conference on logic for programming, artificial intelligence and reasoning, Alicante, Spain, May 22-27, 2020, EasyChair, EPiC Series in Computing, vol 73, pp 354–372. https://doi.org/10.29007/p16v
3. L'Ecuyer P, Le Gland F, Lezaud P, Tuffin B (2009) Splitting techniques. John Wiley & Sons, chap 3:39–61. https://doi.org/10.1002/9780470745403.ch3
4. Budde CE, Ruijters E, Stoelinga MIA (2020) The dynamic fault tree rare event simulator. In: Gribaudo M, Jansen DN, Remke A (eds) Proceedings of the 17th International Conference on Quantitative Evaluation of SysTems (QEST 2020), Springer, Lecture Notes in Computer Science, vol 12289, pp 233–238, https://doi.org/10.1007/978-3-030-59854-9_17

Importance Measures

13

Importance measures, also known as *importance factors*, evaluate the influence of input probabilities on the top event probability [1–3]. In other words, they quantify how changes in a single input probability affect the probability of the top event. These measures offer a quantitative perspective on the key contributors to the system's risk and sensitivity to variations in input parameters: The higher the importance factor of a basic event, the more significant its role in the overall system reliability. Therefore, obtaining accurate data on its exact value is essential.

This chapter discusses six of those measures:

- Structural Importance: $SI(e)$.
- Fussell-Vesely: $FV(e)$.
- Fractional Contribution: $FC(e)$.
- Risk Increase Factor and Risk Increase Interval: $RIF(e)$ and $RII(e)$.
- Risk Decrease Factor and Risk Decrease Interval: $RDF(e)$ and $RDI(e)$.
- Birnbaum Importance: $BI(e)$.

Table 13.1 summarizes the mathematical definitions of these importance measures. Terminology in the literature is not always consistent, with the same terms sometimes having different definitions.

While all measures serve the same purpose, their focus is slightly different: The Structural Importance does not use probabilistic information; the Fussell-Vesely importance is based on cut sets; the Risk Increase and Decrease Factors compare the effect of setting a basic event to 0 or 1; and the Birnbaum Importance uses derivative of the unreliability function. Despite their differences, the probabilistic importance measures SI, FV, FC, RDF, and BI often yield the same ranking of basic events.

An importance analysis typically evaluates several importance measures for all basic events, components, or other relevant objects. Doing so allows the analyst

© The Author(s), under exclusive license to Springer Nature Switzerland AG 2026
M. Stoelinga et al., *Concise Guide to Fault Tree Analysis*, Computer Science Foundations and Applied Logic, https://doi.org/10.1007/978-3-031-78287-9_13

Table 13.1 Summary of definitions of importance measures

Name	Abbreviation		Formula
Structural Importance	$SI(e)$	$=$	$\dfrac{\#\{\mathbf{b} \in \{0, 1\}^n \mid \Phi(\mathbf{b}) = 0 \wedge \Phi(\mathbf{b}[e \leftarrow 1]) = 1\}}{2^{n-1}}$
Fussell-Vesely	$FV(e)$	$=$	$\dfrac{\mathbb{P}[\text{Some MCS } C \text{ containing } e \text{ failed}]}{\mathbb{P}[Top]}$
Risk Decrease Interval	$RDI(e)$	$=$	$\mathbb{P}[Top] - \mathbb{P}[Top \mid \neg e]$
Risk Decrease Factor (Risk Reduction Ratio)	$RDF(e)$	$=$	$\dfrac{\mathbb{P}[Top]}{\mathbb{P}[Top \mid \neg e]}$
Risk Increase Interval	$RII(e)$	$=$	$\mathbb{P}[Top \mid e] - \mathbb{P}[Top]$
Risk Increase Factor (Risk Achievement Worth)	$RIF(e)$	$=$	$\dfrac{\mathbb{P}[Top \mid e]}{\mathbb{P}[Top]}$
Fractional Contribution	$FC(e)$	$=$	$1 - \dfrac{\mathbb{P}[Top \mid \neg e]}{\mathbb{P}[Top]}$
Birnbaum Importance	$BI(e_i)$	$=$	$\dfrac{\delta\, \Phi^{\mathrm{p}}(p_1, \ldots, p_n)}{\delta p_i}$

to rank, compare, and address excessively high importance values. The algorithms to compute the probabilistic importance measures first compute the *nominal value*, which is the top event probability using the original event probabilities $\Phi(\mathbf{p})$. Then the algorithm iterates over each basic event, adjusts them one by one to the desired value, and recalculates the top event probability, e.g., $\Phi(\mathbf{p}[e \leftarrow 0])$, see Table 13.1.

Scope. To simplify notation, we present all definitions for a single basic event. Most importance measures can be extended to groups of basic events, allowing one to assess the importance of entire subsystems or components with multiple failure modes. We also present all definitions in the point probabilistic model. They can also be studied for time-dependent probability distributions, by replacing the point probabilities by the cumulative distributions.

Notation. We assume a fault tree F with basic events $e_1, e_2, \ldots, e_n$ whose probabilities are $\mathbf{p} = (p_1, p_2, \ldots, p_n)$. We write L for the set of all cut sets of F. Recall that we use expressions 'an event e fails' and 'an event e occurs' as synonyms.

Recall that $\Phi_F^{\mathrm{p}}(\mathbf{p})$ is the probabilistic structure function, yielding the top event probability, given the input probabilities $\mathbf{p}$. We often abbreviate both Φ_F and Φ_F^{p} by Φ.

For an event e we write $\mathbf{p}[e \leftarrow x]$ for the probability vector where the probability of the event e is set to x. If e is the event with the index i, then $\mathbf{p}[e \leftarrow x]$ is the same vector as $\mathbf{p}$ except that $p_i = x$. The top event probability with the modified value of e is denoted by $\Phi(\mathbf{p}[e \leftarrow 0])$.

All probabilistic importance measures depend on F and on $\mathbf{p}$. The latter dependence is not explicit in the notation, so $FV(e)$ should be read as $FV_F(e, \mathbf{p})$, and the same for the other measures.

13.1 Structural Importance

The *Structural Importance*, abbreviated *SI*, evaluates the impact of the place of a basic event in the system. Its definition does not use probabilistic information, and can therefore be used as a first assessment at early design stages, when little information on the components' reliability is available. Structural Importance measures how often a basic event e is critical to system failure. Specifically, it counts the number of status vectors where changing e from 0 (working) to 1 (failed) causes the top event to change from 0 to 1, indicating a system failure.

Definition 16 A basic event e is *critical* in a status vector $\mathbf{b}$ if its failure will cause a system failure.

$$\Phi(\mathbf{b}) = 0 \quad \text{and} \quad \Phi(\mathbf{b}[e \leftarrow 1]) = 1.$$

Here, $\mathbf{b}[e \leftarrow 1]$ is the status vector that is the same as $\mathbf{b}$, except that for $e_i = e$, the status bit b_i has been set to 1.

The Structural Importance of basic event e counts the number of status vectors where e is critical, as the fraction of all status vectors in which e has failed.

$$SI(e) = \frac{\#\{\mathbf{b} \in \{0, 1\}^n \mid e \text{ is critical in } \mathbf{b}\}}{2^{n-1}}$$

Note that, if e_i is critical in $\mathbf{b}$, then b_i must be 0. Therefore, the denominator is 2^{n-1}, i.e., the number of status vectors with $b_i = 0$.

An equivalent definition looks at criticality in the opposite direction: By counting the number of status vectors where the fault tree has failed, and would be functional if e were not failed:

$$SI(e) = \frac{\#\{\mathbf{b} \in \{0, 1\}^n \mid \Phi(\mathbf{b}) = 1 \wedge \Phi(\mathbf{b}[e \leftarrow 0]) = 0\}}{2^{n-1}}$$

The higher the Structural Importance of a basic event, the more critical it is for the top failure: if $SI(e)$ is high, then there are many scenarios (i.e., status vectors) where the failure of e makes the top event fail.

Example 34

Consider the basic event e_1 in the fault tree $F = \text{AND}(e_1, e_2, e_3, e_4)$. Then e_1 is critical in only one status vector, namely 0111. Thus, we have $SI(e_1) = \frac{1}{8}$. Consider e_1 in $\text{OR}(e_1, e_2, e_3, e_n)$. Again, e_1 is critical in a single status vector, namely 0000, yielding $SI(e_1) = \frac{1}{8}$. In both cases, all basic events have the same Structural Importance.

In $\text{OR}(\text{AND}(e_1, e_2), e_3, e_4)$, event e_1 is critical in 0100, while e_3 is critical in 0000, 1000, and 0100. This sets $SI(e_1) = \frac{1}{8}$ and $SI(e_3) = \frac{3}{8}$. This is reasonable, because e_1 and e_2 can be seen as redundant components, so their failure is less critical than the failure of non-redundant components e_3 and e_4.

In $\text{AND}((\text{OR}(e_1, e_2), e_3, e_4)$, we see that e_1 is critical only in 0011. Thus, $SI(e_1) = \frac{1}{8}$. However, e_3 is critical in 0101, 1001, and 1101. Thus, $SI(e_3) = \frac{3}{8}$. This is again reasonable: If e_1 fails, then two specific failures need to occur, namely e_3 and e_4, for the top event to fail. If e_3 fails, then a top failure needs e_4 to fail, and either e_1 or e_2 or both. Incorporating more failure opportunities, e_3 has a higher Structural Importance than e_1.

13.2 Fussell-Vesely Importance Measure

The *Fussell-Vesely* importance, abbreviated FV, measures the contribution of all minimal cut sets containing the basic event e to the total failure probability. That is, it divides the probability that some MCS containing e fails by the top probability.

Definition 17 Let e be a basic event. Then the Fussell-Vesely importance $FV(e)$ is defined by

$$FV(e) = \frac{\mathbb{P}[\text{Some MCS } C \text{ containing } e \text{ failed}]}{\mathbb{P}[Top]}$$
$$= \frac{\mathbb{P}[\bigvee_{C \in L, e \in C} C]}{\mathbb{P}[Top]}$$

The FV importance can be understood as the probability that e contributed to the failure of the top event, given that the top failed. In order for e to contribute to the top failure, some cut set containing e must fail.

Table 13.2 Cut set probabilities. We write T5 for Sp. Let $\mathbb{P}[L]$ denote the sum of all minimal cut set probabilities in the road trip example. There are 10 minimal cut sets of the form Con, Ti, Tj, so we count their probability 10 times.

Basic event	Full name	Probability	
Con	No connection	0.25	
Bat	Battery fails	0.2	
Eng	Engine fails	0.05	
Ti	Tire i fails	0.1	
Minimal cut set	**Probability**		
Con, Eng	$0.25 \cdot 0.05$	$= 0.0125$	$= 1.25\text{E-}2$
Bat, Eng	$0.2 \cdot 0.05$	$= 0.01$	$= 1.0\text{E-}2$
Con, Ti, Tj	$10 \cdot 0.25 \cdot 0.1 \cdot 0.1$	$= 0.025$	$= 2.5\text{E-}2$
Total	$\mathbb{P}[L]$	$= 0.0475$	$= 4.75\text{E-}2$

$$\mathbb{P}[\text{Some MCS } C \text{ containing } e \text{ failed} \mid \text{top fails}]$$

$$= \frac{\mathbb{P}[\text{Some MCS } C \text{ containing } e \text{ failed} \wedge \text{top fails}]}{\mathbb{P}[\text{top fails}]}$$

$$= \frac{\mathbb{P}[\text{Some MCS } C \text{ containing } e \text{ failed}]}{\mathbb{P}[\text{top fails}]}$$

$$= FV(e)$$

Computation of FV. The FV importance is often, but not always, computed by approximating the top probability $\mathbb{P}[Top]$ via the rare event and min cut upper bound approximations:

$$FV(e) \leq \frac{1 - \prod_{C \in L, e \in C}(1 - \mathbb{P}[C])}{\mathbb{P}[Top]}$$

$$\leq \frac{\sum_{C \in L, e \in C} \mathbb{P}[C]}{\mathbb{P}[Top]}$$

Here L denotes the set of all minimal cut sets in the fault tree. Sometimes, $\mathbb{P}[Top]$ is also approximated by the cut set probabilities. However, this gives no bounds on the FV importance. The reason is that $\mathbb{P}[\bigvee_{C \in L, e \in C} C] \leq \sum_{C \in L, e \in C} \mathbb{P}[C]$, while $\mathbb{P}[Top] \leq \sum_{C \in L} \mathbb{P}[C]$, and similarly for the min cut upper bound approximation.

Example 35

We compute the FV importance factor for several BEs in the road trip example. First, we recall the minimal cut set probabilities from Example 2, we repeat these probabilities in Table 13.2.

Example 35a *Battery.* To obtain the FV importance for the battery, we observe that the battery only appears in a single minimal cut set {Bat, Eng}, which has probability 0.01. Thus, we obtain:

$$FV(\text{Bat}) = \frac{\mathbb{P}[\{\text{Bat, Eng}\}]}{\mathbb{P}[L]} = \frac{0.01}{0.0475} \approx 0.21$$

This means that around 21% of the total cut set probabilities can be attributed to a cut set where the battery is involved.

Example 35b *Engine.* The engine appears in two minimal cut sets, {Con, Eng} and {Bat, Eng}, with probabilities 0.0125 and 0.01. Thus, we get

$$FV(\text{Eng}) = \frac{\mathbb{P}[\{\text{Con, Eng}\} + \{\text{Bat, Eng}\}]}{\mathbb{P}[L]} = \frac{0.0225}{0.0475} \approx 0.473.$$

Thus, around 47% of the total cut set probability involves a cut set containing the engine.

Note that if some MCS containing e fails, then both e and top event fails. The converse is not true: If e fails, no cut set with e needs to fail; and if the top event fails, no cut set involving e has to fail.

13.3   Fractional Contribution Importance Measure

The *Fractional Contribution*, abbreviated *FC*, of a basic event gives the contribution of this basic event to the top event probability. The *FC* importance value yields the fraction of the top value that disappears from the top value if we decrease the event failure probability to zero. Under the assumption that the system failure probability is linear in the event probability, it is also the fraction that is added to the top value if the failure probability of the event doubles.

For example, if the Fractional Contribution of a basic event e in a fault tree F with a top event *Top* is 0.3 then, intuitively, 30% of the top event probability $\mathbb{P}[Top]$ are caused by e. If e never occurs, then the top event probability changes to $\mathbb{P}[Top] - \mathbb{P}[Top] \cdot 0.3$, because e stops to contribute to the failure probability of F. If the failure probability of e doubles or if we identify another failure source with the same logical position as e and the same failure probability and add it to the model, then the new top event probability will be $\mathbb{P}[Top] + \mathbb{P}[Top] \cdot 0.3$.

Table 13.3 Fractional Contribution values for basic events in the road trip example

Basic event	$\Phi(\mathbf{p}[a \leftarrow 0])$	FC
Con	0.01	0.746
Eng	0.0204	0.482
Bat	0.0318	0.191
Ti	0.0324	0.176

Definition 18 Let e be a basic event. Then the Fractional Contribution of e, denoted $FC(e)$, is defined by

$$FC(e) = 1 - \frac{\mathbb{P}[Top \mid e \text{ does not occur}]}{\mathbb{P}[Top]}$$
$$= 1 - \frac{\Phi(\mathbf{p}[e \leftarrow 0])}{\Phi(\mathbf{p})}$$

Computation of FC. The definition requires a quantification of the top event value with the updated failure probability of the event e. Any quantification method presented in this book allows this with the same complexity as the quantification of the top event probability. If we have calculated a list of minimal cut sets for the fault tree F then we can use this list to approximate $\Phi(\mathbf{p}[e \leftarrow 0])$. In this particular case where we set the probability of e to zero, we can delete minimal cut sets that contain e and quantify the minimal cut set list consisting of the remaining cut sets. We can use the same algorithms as those described in Chap. 8. Quantification based on Binary Decision Diagrams (BDDs) can handle updated probabilities equally well. The same quantification algorithm can be used for importance analyses. The same is true also for a BDD built from a minimal cut set list.

Example 36

Let us look at the road trip example. The top event probability calculation with original event probabilities gives us $\Phi(\mathbf{p}) = 0.0393$. Fractional Contribution values are given in Table 13.3.

The linearity assumption is valid for all of these events. For example, if we multiply the failure probability of any of the events by two, then the top value increases by the top event probability multiplied by the Fractional Contribution. If we multiply the event probability by three, it increases again by the same amount.

We can also see that almost three quarters of the top event probability are driven by the connection failure. If we can decrease its failure probability to one half then we improve the system more than by decreasing the failure probability of the engine or of the battery or of a tire to one half.

13.4   Risk Increase Importance

The Risk Increase importance of a basic event calculates the amount that the total probability would increase if that basic event would be unavailable. It is especially relevant for assessing the situation when equipment is taken out of operation, for example, due to service or maintenance.

There are two variants of the Risk Increase importance. The *Risk Increase Interval* $RII(e)$ treats the risk increase as an absolute value, i.e., it takes the difference with the situation where the component can fail. The *Risk Increase Factor RIF(e)* treats it as a value relative to the top probability. This measure is also called the *Risk Achievement Worth (RAW)*, as it quantifies the worth of a component in achieving the present level of reliability. Both can be expressed via the probabilistic structure function.

Definition 19 Let e be a basic event. Then the Risk Increase Interval $RII(e)$ and the Risk Increase Factor $RIF(e)$ are defined by

$$RII(e) = \mathbb{P}[Top \text{ occurs} \mid e \text{ always occurs}] - \mathbb{P}[Top]$$
$$= \Phi(\mathbf{p}, p[e \leftarrow 1]) - \Phi(\mathbf{p})$$

$$RIF(e) = \frac{\mathbb{P}[Top \text{ occurs} \mid e \text{ always occurs}]}{\mathbb{P}[Top]}$$
$$= \frac{\Phi(\mathbf{p}, p[e \leftarrow 1])}{\Phi(\mathbf{p})}$$

Computing *RII* and *RIF*. Several methods exist to compute the Risk Increase importance. For an exact result, one needs to re-evaluate the fault tree, where the event under consideration is set to 1. This means modifying the original fault tree, generating minimal cut sets or building a BDD, and finally calculating the top value. This is a time consuming process. Typically, software tools use already calculated results, either a minimal cut set list or a BDD. Setting the probability of a basic event to 1 and re-calculating the top value is a fast operation, irrespective of whether we use a BDD or a minimal cut set list.

This approach requires additional care. Application of cutoff discards parts of the solution which might be relevant for the risk increase. Consider a situation where all minimal cut sets that contain an event e fall below the cutoff. The Risk Increase Factor of e equals 1. If we considered also minimal cut sets with e included, the Risk Increase Factor could be significantly higher, especially if e has a very low failure probability.

Additionally, setting the probability of an event to 1 means that we remove it from the cut set list. This might render some cut sets non-minimal and lead to overly conservative Risk Increase Factor values, especially if we calculate the Risk Increase Factor for several basic events at once. One option to avoid overly conservative values is to minimize the cut set list after removing the event e from the cut sets. This issue

Table 13.4 Cut set probabilities with the original event probabilities and when Bat is assumed to always fail

Minimal cut set	Probability			Probability	If Bat fails
Con, Eng	$0.25 \cdot 0.05$	$= 0.0125$			0.0125
Bat, Eng	$0.2 \cdot 0.05$	$= 0.01$			0.05
Con, Ti, Tj	$10 \cdot 0.25 \cdot 0.1 \cdot 0.1$	$= 0.025$			0.025
Total	$\mathbb{P}[L]$	$= 0.0475$		$\mathbb{P}[L'] =$	0.0875

We get that $RIF(\text{Bat}) = \mathbb{P}[L']/\mathbb{P}[L] = 1.842$. But we can observe that the first cut set is not minimal anymore. If we remove it from the cut set list and re-calculate $\mathbb{P}[L''] = 0.075$, then we obtain $RIF(\text{Bat}) = 1.579$ instead

Table 13.5   Exact Risk Increase Factor values for basic events in the road trip example

Basic event	$\Phi(\mathbf{p}[a \leftarrow 1])$	RIF
Eng	0.4	10.166
Con	0.127	3.238
Ti	0.102	2.584
Bat	0.069	1.763

The exact $RIF(\text{Bat})$ value is lower than the one estimated by the Rare Event Approximation without minimization, but higher than the one with the additional minimization step. This is caused by the conservatism in the Rare Event Approximation of the original top event probability

dissolves if we quantify the minimal cut set list by a BDD or if we use the BDD quantification for the whole fault tree.

The definition of the Risk Increase Factor implicitly assumes that the complete unavailability of a component is expressed by setting its value to 1 (or, by removing it). Frequency basic events, typically used as accident initiating events, do not have this property. It is not possible to calculate Risk Increase importance for them.

> **Example 37**
>
> Table 13.4 calculates the Risk Increase Factor for the event Bat in the road trip example by updating minimal cut set probabilities and calculating the top event probability by the Rare Event Approximation. Exact Risk Increase Factor values for all basic events are given in Table 13.5.

13.5   Risk Decrease Importance (Improvement Potential)

The Risk Decrease or Risk Reduction importance is similar to the Risk Increase importance from the previous section, except that it considers the situation where the basic event never fails. Again two variants exist. The *Risk Decrease Interval*

Table 13.6 Cut set probabilities with the original event probabilities and when Bat is assumed to never fail

Minimal cut set	Probability		Probability if Bat never fails
Con, Eng	$0.25 \cdot 0.05$	$= 0.0125$	0.0125
Bat, Eng	$0.2 \cdot 0.05$	$= 0.01$	0
Con, Ti, Tj	$10 \cdot 0.25 \cdot 0.1 \cdot 0.1$	$= 0.025$	0.025
Total	$\mathbb{P}[L]$	$= 0.0475$	$\mathbb{P}[L'] = 0.0375$

We obtain $RDF(\text{Bat}) = \mathbb{P}[L]/\mathbb{P}[L'] = 1.267$

$RDI(e)$ treats the risk decrease as a difference, and the *Risk Decrease Factor* (or, *Risk Reduction Ratio*) $RDF(e)$ treats it as a ratio.

Definition 20 Let e be a basic event. Then the Risk Decrease Interval $RDI(e)$ and the Risk Decrease Factor, sometimes called Risk Reduction Ratio, $RDF(e)$ are respectively defined by

$$RDI(e) = \mathbb{P}[Top] - \mathbb{P}[Top \mid e \text{ does not occur}]$$
$$= \Phi(\mathbf{p}) - \Phi(\mathbf{p}, p[e \leftarrow 0])$$

$$RDF(e) = \frac{\mathbb{P}[Top]}{\mathbb{P}[Top \mid e \text{ does not occur}]}$$
$$= \frac{\Phi(\mathbf{p})}{\Phi(\mathbf{p}, p[e \leftarrow 0])}$$

It can be shown that these two measures give exactly the same ranking as the Fractional Contribution measure.

Computation of RDI and RDF. Similar to *RII* and *RIF*.

> *Example 38*
>
> Table 13.6 calculates the Risk Decrease Factor for the event **Bat** in the road trip example by updating minimal cut set probabilities and calculating the top event probability by the Rare Event Approximation. Exact Risk Decrease Factor values for all basic events are given in Table 13.7.

Table 13.7   Exact Risk Decrease Factor values for basic events in the road trip example

Basic event	$\Phi(\mathbf{p}[a \leftarrow 0])$	RDF
Con	0.01	3.935
Eng	0.0204	1.932
Bat	0.0319	1.236
Ti	0.0324	1.214

The exact *RDF*(Bat) value is lower than the one estimated by the Rare Event Approximation

Relation to Structural Importance. The Risk Decrease Interval can be seen as a generalization of the Structural Importance when probabilities are considered. The two measures coincide (except for a constant factor 2) when all basic event probabilities are 1/2: In this case, we note that each status vector $\mathbf{b}$ of an n-event FT has equal probability $P[\mathbf{b}] = 2^{-n}$. This gives rise to the following derivation:

$$
\begin{aligned}
SI(e) &= \frac{\#\{\mathbf{b} \in \mathbb{B}^n \mid \Phi(\mathbf{b}) = 1 \wedge \Phi(\mathbf{b}[e \leftarrow 0]) = 0\}}{2^{n-1}} \\
&= \frac{\#\{\mathbf{b} \in \mathbb{B}^n \mid \Phi(\mathbf{b}) = 1\} - \#\{\mathbf{b} \in \mathbb{B}^n \mid \Phi(\mathbf{b}) = 1 \wedge \Phi(\mathbf{b}[e \leftarrow 0]) = 1\}}{2^{n-1}} \\
&= \frac{\#\{\mathbf{b} \in \mathbb{B}^n \mid \Phi(\mathbf{b}) = 1\} - \#\{\mathbf{b} \in \mathbb{B}^n \mid \Phi(\mathbf{b}[e \leftarrow 0]) = 1\}}{2^{n-1}} \qquad \Phi \text{ monotonic} \\
&= \left(2^{-n+1} \sum_{\mathbf{b} \in \mathbb{B}^n} \Phi(\mathbf{b})\right) - \left(2^{-n+1} \sum_{\mathbf{b} \in \mathbb{B}^n} \Phi(\mathbf{b}[e \leftarrow 0])\right) \\
&= \left(2 \sum_{\mathbf{b} \in \mathbb{B}^n} \Phi(\mathbf{b}) \cdot \mathbb{P}[\mathbf{b}]\right) - \left(2 \sum_{\mathbf{b} \in \mathbb{B}^n} \Phi(\mathbf{b}[e \leftarrow 0]) \cdot \mathbb{P}[\mathbf{b}]\right) \\
&= 2\mathbb{P}[Top] - \left(2 \sum_{\mathbf{b} \in \mathbb{B}^n} \Phi(\mathbf{b}[e \leftarrow 0]) \cdot \mathbb{P}[\mathbf{b}]\right) \\
&= 2\mathbb{P}[Top] - \left(2 \sum_{\mathbf{b} \in \mathbb{B}^n} \Phi(\mathbf{b}) \cdot (1 - \Phi_e(\mathbf{b})) \cdot 2 \cdot \mathbb{P}[\mathbf{b}]\right) \\
&= 2\mathbb{P}[Top] - 2\mathbb{P}[Top \wedge \neg e] \cdot 2 \\
&= 2\mathbb{P}[Top] - 2\frac{\mathbb{P}[Top \wedge \neg e]}{\mathbb{P}[\neg e]} \qquad \text{since } \mathbb{P}[\neg e] = \frac{1}{2} \\
&= 2(\mathbb{P}[Top] - \mathbb{P}[Top \mid \neg e]) = 2 \cdot RDI(e)
\end{aligned}
$$

13.6   Birnbaum Importance Measure

The Birnbaum Importance measure (BI) indicates the rate of change in the top probability as a result of changes in the basic event probabilities. Thus, the BI importance of a basic event e_i is defined as the partial derivative of the top probability with respect to p_i. Recall that Φ^P is the probabilistic structure function, yielding $p_{Top} = \Phi^P(p_1, \ldots, p_n)$.

Definition 21 The Birnbaum Importance measure for basic event e_i is given by

$$BI(e_i) = \frac{\delta\, \Phi^P(p_1, \ldots, p_n)}{\delta p_i}$$

Birnbaum Importance yields the sensitivity of p_{Top} in p_e. The higher $BI(e)$, the more e matters for the system reliability.

Computing BI. The Birnbaum Importance can be computed in multiple ways. If we have an expression for the structure function Φ^P, then we can apply Definition 21 directly and take its derivative, obtaining an exact result for $BI(e)$. Recall that Φ^P is always a polynomial, so finding its derivative is conceptually not difficult. An expression for Φ^P can also be approximated via the minimal cut sets, yielding an approximate result for $BI(e)$. To do so, note that the sum of all cut set probabilities $\mathbb{P}[L] = \sum_{C \in L} \prod_{c \in C} p_c$ yields a symbolic expression in $p_1, \ldots, p_n$. This latter also works if the list of minimal cut sets is approximated, e.g., via truncation. Finally, as for any partial derivative, we can approximate $\Phi^{P'}$ numerically. Then we pick a small h and set

$$BI(e_i) \approx \frac{\Phi^P((p_1, \ldots, p_n)[i \leftarrow p_i + h]) - \Phi^P(p_1, \ldots, p_n)}{h}$$

Here, $(p_1, \ldots, p_n)[i \leftarrow p_i + h]$ is the same vector as $(p_1, \ldots, p_n)$ except that it contains the value $p_i := p_i + h$ at position i.

Example 39

Consider the fault tree below. Its Boolean and probabilistic structure function are respectively given by

$$\Phi(\text{Con, Bat, Eng}) = \text{Con} \ \vee \ (p_{\text{Bat}} \wedge p_{\text{Eng}})$$

$$\Phi^P(p_{\text{Con}}, p_{\text{Bat}}, p_{\text{Eng}}) = p_{\text{Con}} + p_{\text{Bat}} p_{\text{Eng}} - p_{\text{Con}} p_{\text{Bat}} p_{\text{Eng}}$$

Thus, we obtain the Birnbaum Importance of Con as

$$
\begin{aligned}
BI(\text{Con}) &= \frac{\delta \ \Phi^P(p_{\text{Con}}, p_{\text{Bat}}, p_{\text{Eng}})}{\delta p_{\text{Con}}} \\
&= \frac{\delta(p_{\text{Con}} + p_{\text{Bat}} p_{\text{Eng}} - p_{\text{Con}} p_{\text{Bat}} p_{\text{Eng}})}{\delta p_{\text{Con}}} \\
&= 1 - p_{\text{Bat}} p_{\text{Eng}} \\
&= 1 - 0.2 \cdot 0.05 = 0.99
\end{aligned}
$$

The expression $BI(\text{Con}) = 1 - p_{\text{Bat}} \cdot p_{\text{Eng}}$ shows that p_{Bat} and p_{Eng} are very high, then $BI(\text{Con})$ is low, so the dependence on Eng is low. This is reasonable, because in that case, the phone will already fail due to the battery and connection (Fig. 13.1).

Fig. 13.1 Phone failure

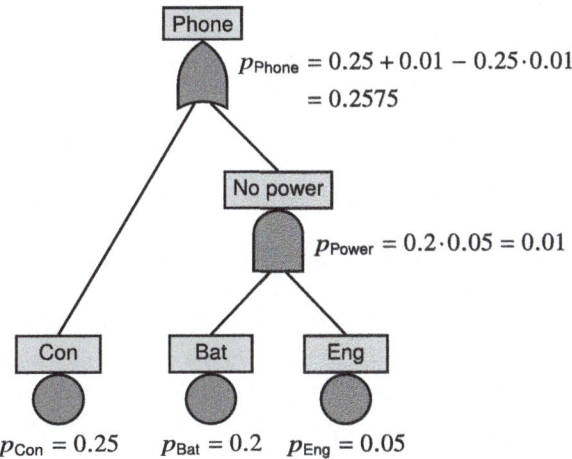

$$p_{\text{Phone}} = 0.25 + 0.01 - 0.25 \cdot 0.01$$
$$= 0.2575$$

$$p_{\text{Power}} = 0.2 \cdot 0.05 = 0.01$$

$$p_{\text{Con}} = 0.25 \qquad p_{\text{Bat}} = 0.2 \qquad p_{\text{Eng}} = 0.05$$

BI via Minimal Cut Set Probabilities. Using the minimal cut set probabilities, yields the following. There are two minimal cut sets, {Con} and{Bat, Eng}, therefore

$$\Phi^{MCS}(p_{Con}, p_{Bat}, p_{Eng}) = p_{Con} + p_{Bat} \cdot p_{Eng}$$

Using Φ^{MCS} for the Birnbaum Importance yields

$$
\begin{aligned}
BI(Con) &= \frac{\delta\,\Phi^{MCS}(p_{Con}, p_{Bat}, p_{Eng})}{\delta p_{Con}} \\
&= \frac{\delta(p_{Con} + p_{Bat} \cdot p_{Eng})}{\delta p_{Con}} = 1
\end{aligned}
$$

This is a good approximation when p_{Bat} and p_{Eng} are low. In that case, the failure of Phone fails is dominated by p_{Con}, if we increase p_{Con}, then $p_{Phone\ fails}$ increases at the same rate.

BI via Numerical Differentiation. Finally, we illustrate how to approximate $BI(Con)$ by numerically approximating the derivative. We take $h = 0.001$. Then

$$
\begin{aligned}
BI(Con) &\approx \frac{\Phi^P(p_{Con}, p_{Bat}, p_{Eng}) - \Phi^P(p_{Con} + h, p_{Bat}, p_{Eng})}{h} \\
&= \frac{(p_{Con} + h) + p_{Bat}p_{Eng} - (p_{Con} + h)p_{Bat}p_{Eng} - (p_{Con} + p_{Bat}p_{Eng} - p_{Con}p_{Bat}p_{Eng})}{h} \\
&= \frac{h + hp_{Bat}p_{Eng}}{h} \\
&\quad 1 + p_{Bat}p_{Eng} = 1 + 0.2 \cdot 0.05 = 1.01
\end{aligned}
$$

13.7 Relations Between Importance Measures

Importance measures rank basic events or components according to the role they play in the system risk. Different aspects of this risk require different importance measures. A good understanding of the differences and similarities between importance measures helps us to select the correct measure for the purpose that we have in mind. Table 13.8 summarizes relations between importance measures.

Two aspects define the dependability contribution of a component in a system:

- Dependability of the component itself.
- Position of the component in the system design.

We also distinguish the importance measures based on whether they cover one or both of these contribution sources.

Table 13.8 Relations between importance measures

Importance	Relation	Importance
FC	$FC(e) \approx FV(e)$	
	$FC = FV$ if $\mathbb{P}[Top] - \mathbb{P}[Top \mid e$ does not occur$]$ is equal to $\mathbb{P}[$Some MCS C containing e failed$]$	FV
	$FC(e) = 1 - \dfrac{1}{RDF(e)}$	RDF
	$FC(e) = \dfrac{RDI(e)}{\mathbb{P}[Top]}$	RDI
BI	$BI(e) = \dfrac{FC(e) \cdot \mathbb{P}[Top]}{\mathbb{P}[e]}$	FC
	$BI(e) \approx \mathbb{P}[Top \mid e$ does not occur$] \cdot (RIF(e) - 1)$	RIF
	$BI(e) = RII(e) + RDI(e)$	RII, RDI

Fractional Contribution. Let us first look at the Fractional Contribution measure. It tells us what effects improvements of the component failure probability would have on the top event probability. Alternatively, instead of changing the component characteristics, one could also re-design the system so that this component plays a different role. One could add a redundant subsystem with a certain failure probability. This decreases the top event probability, because it decreases the failure probability of the original component by the factor equal to the failure probability of the redundant subsystem, assuming there are no common cause failures involved. The higher the Fractional Contribution factor is, the greater the potential for dependability improvements.

Fractional Contribution is related to the Fussell-Vesely measure and to the Risk Decrease Factor. Let us first compare definitions of Fractional Contribution and Fussell-Vesely.

$$FC(e) = 1 - \frac{\mathbb{P}[Top \mid e \text{ does not occur}]}{\mathbb{P}[Top]} = \frac{\mathbb{P}[Top] - \mathbb{P}[Top \mid e \text{ does not occur}]}{\mathbb{P}[Top]}$$

$$FV(e) = \frac{\mathbb{P}[\text{Some MCS } C \text{ containing } e \text{ failed}]}{\mathbb{P}[Top]}$$

We can see that they coincide if $\mathbb{P}[Top] - \mathbb{P}[Top \mid e$ does not occur$]$ is equal to $\mathbb{P}[$Some MCS C containing e failed$]$. The probability $\mathbb{P}[Top \mid e$ does not occur$]$ can be characterized by minimal cut sets that do not contain e. On the first look, subtracting this probability from the top event probability is equivalent to calculating $\mathbb{P}[$Some MCS C containing e failed$]$. But this holds only if we use the rare event approximation to calculate probabilities. Otherwise, once we use a more precise method to estimate probabilities, these two measures do not coincide and the difference grows with the conservatism in the rare event approximation. If these two values differ, the Fussell-Vesely measure should be interpreted as an approximation of the fractional contribution.

An alternative definition of Fussell-Vesely would re-establish the equivalence for any quantification method:

$$FV(e) = \frac{\mathbb{P}[\textit{Top}] - \mathbb{P}[\text{Some MCS } C \text{ that does not contain } e \text{ failed}]}{\mathbb{P}[\textit{Top}]}$$

The conceptual and numerical similarity of Fractional Contribution and Fussell-Vesely result in some literature, e.g., [1,4], using the name Fussell-Vesely for Fractional Contribution.

Another importance measure closely related to Fractional Contribution is the Risk Decrease importance. Straight from the definitions, one can establish the following equalities.

$$FC(e) = 1 - \frac{\mathbb{P}[\textit{Top} \mid e \text{ does not occur}]}{\mathbb{P}[\textit{Top}]} = 1 - \frac{1}{RDF(e)}$$

$$FC(e) = \frac{RDI(e)}{\mathbb{P}[\textit{Top}]}$$

Both importance measures produce the same ordering of basic events, but the direct interpretations differ. The Risk Decrease Factor tells how much safer the whole system becomes when we make this basic event never happen, i.e., the component is perfect, at least with respect to the failure mode corresponding to this basic event.

Fractional Contribution, Fussell-Vesely, and Risk Decrease Factor belong to the category of *design* importance measures. They inform analysts about risk-based priorities when trying to improve the system design. All of them cover both aspects of the component contribution to the system dependability. The dependability of the component and its position in the system design structure affect the numerical value of these measures.

Birnbaum Importance. One might want to understand the penalty in terms of the increase of the top event probability if, say for cost reasons, the dependability of the component decreases. Or, if we use a copy of this component at other places which are serially connected to the original one, which might double the number of related failure scenarios. We obtain those that involve the original component and symmetrical ones that contain the copy one.

The best relaxation of component dependability properties affects the overall system dependability only minimally. The Birnbaum Importance measure covers only one aspect of the component contribution to the system dependability related to the position of the component in the system design structure. It is completely independent of the failure probability of the component. Typically, this is sufficient to select the most suitable components for relaxation, namely those with the lowest Birnbaum Importance measure.

Fractional Contribution can increase the confidence in the choice of components by supplying information about the component itself. High Fractional Contribution should raise a flag and lead to a closer inspection, because it sends a signal that improving component dependability has a good potential to improve the system. Low Fractional Contribution confirms the indication of the Birnbaum Importance measure.

The relations to Fractional Contribution and the Risk Increase Factor are given below.

$$BI(e) = \frac{FC(e) \cdot \mathbb{P}[Top]}{\mathbb{P}[e]}$$

$$BI(e) \approx \mathbb{P}[Top \mid e \text{ does not occur}] \cdot (RIF(e) - 1)$$

Risk Increase Factor. This importance measure instantiates the relaxation question for the most extreme case—the component always fails. We can also say that the component is totally unavailable. For reliable components, Risk Increase Factor, similarly to the Birnbaum Importance, only covers the position of the component in the system design structure. It answers the question about how well the rest of the system can handle the absence of this component.

Risk Increase Factor belongs to the category of *configuration* importance measures. It warns us for certain configuration changes that could compromise the system dependability. We should not take components with a high Risk Increase Factor out of operation, e.g., for preventive maintenance. Also, we should not stay in configurations where some components have a high Risk Increase Factor unnecessarily long, because their failure would have significant effect on the system failure probability.

References

1. Borst M, Schoonakker H (2001) An overview of PSA importance measures. Reliabil Eng Syst Safety 72(3):241–245. https://doi.org/10.1016/S0951-8320(01)00007-2
2. Dutuit Y, Rauzy A (2014) Importance factors of coherent systems: a review. J Risk Reliabil 228:313–323. https://doi.org/10.1177/1748006X13512296
3. Kuo W, Zhu X (2012) Importance measures in reliability, risk, and optimization: principles and applications. John Wiley & Sons Ltd. https://doi.org/10.1002/9781118314593
4. Vesely W, Davis T, Denning R, Saltos N (1983) Measures of risk importance and their applications. U.S. Nuclear Regulatory Commission, NUREG/CR–3385

Part V
Extensions

Common Cause Failures

14

Common cause failures (CCFs) are factors that cause failures in multiple components at the same time. CCFs are of crucial concern for risk analysts, as they often dominate the system dependability. Two important types of CCFs exist. *External shocks* are events that affect all or multiple components at the same time, for instance a fire, falling debris, or power outage. In the railroad example, nails on the road present a typical shock common cause. A second type is *common mode failures,* where the underlying condition shared by these components increases their failure probability or rate. Examples are production failures, harsh conditions, such as high-temperature or salted water, and maintenance errors.

Common cause failures break the independence assumption between basic events: if one basic event fails due to a common cause, then all events subject to the same common cause are also more likely to fail. This means that ignoring common cause failures will yield a serious overoptimistic assessment of the system dependability, for instance, by overestimating the positive impact of redundancies.

Therefore, additional methods are required to treat common cause failures in fault trees faithfully. The standard method creates new basic events, the so-called *common cause failure (CCF) events*. These CCF events represent (or include) failures of basic events that share a common failure cause. The set of basic events affected by the same common source of failures is called a *CCF group*.

Additionally, it is important to split the probability of failure of the original event into the probability of a common cause failure and the probability of an independent failure. Significant effort is invested in estimating the parameters of these CCF models, drawing on operational experience and expert judgment [1–3]. The values of these parameters are regularly updated and documented in reference books that list commonly used components within specific industries.

© The Author(s), under exclusive license to Springer Nature Switzerland AG 2026   183
M. Stoelinga et al., *Concise Guide to Fault Tree Analysis*, Computer Science Foundations
and Applied Logic, https://doi.org/10.1007/978-3-031-78287-9_14

In this chapter, we discuss three CCF models and their failure quantification. The β-*factor* model assumes that, if the CCF event occurs, then all elements in the CCF group also fail. The α-*factor* and *multiple Greek letter* model allow subsets of the CCF group to fail by a common cause.

14.1 Dependence

Definition 22 Two events E_1 and E_2 are (probabilistically) independent (a.k.a. statistically or stochastically independent) if

$$\mathbb{P}[E_1 \wedge E_2] = \mathbb{P}[E_1] \cdot \mathbb{P}[E_2]$$

Independence means that knowledge about the occurrence of E_2 does not influence the probability of E_1 and vice versa: $\mathbb{P}[E_1 \mid E_2] = \mathbb{P}[E_1]$ and $\mathbb{P}[E_2 \mid E_1] = \mathbb{P}[E_2]$.
Events E_1 and E_2 have *positive dependence* if

$$\mathbb{P}[E_1 \wedge E_2] > \mathbb{P}[E_1] \cdot \mathbb{P}[E_2]$$

In particular, the occurrence of E_2 increases the probability of occurrence of E_1, i.e., $\mathbb{P}[E_1 \mid E_2] > \mathbb{P}[E_1]$. Vice versa, the occurrence of E_1 increases the probability of E_2, i.e., $\mathbb{P}[E_2 \mid E_1] > \mathbb{P}[E_2]$. They have *negative dependence* if

$$\mathbb{P}[E_1 \wedge E_2] < \mathbb{P}[E_1] \cdot \mathbb{P}[E_2]$$

Then, the occurrence of E_2 decreases the probability of occurrence of E_1, i.e., $\mathbb{P}[E_1 \mid E_2] < \mathbb{P}[E_1]$ and vice versa.
It is especially the positive dependence that is of concern for risk analysis, since that yields an unrealistically optimistic assessment of failure probability, thereby underestimating failure risks.

14.2 Common Cause Failure

Fault tree analysis usually studies dependencies between basic events that arise from a shared cause that affects multiple basic events at the same time.

Definition 23 A set of basic events that are subject to the same cause is called a *CCF group*.

Example 40 (Common cause failure modeled in the β model)

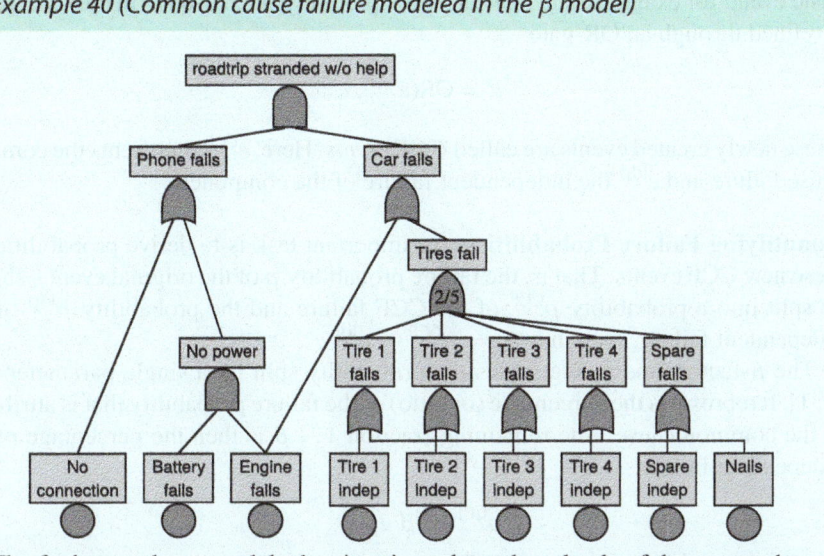

The fault tree above models the situation where the wheels of the car can be subject to a common cause, namely nails on the road. Thus, we define a CCF group consisting of all wheels T1, T2, T3, T4, and Sp. The CCF event Nails models a common cause failure of all tires.

In the figure above, the CCF is modeled explicitly in the fault tree. As shown, failure of each individual tire Ti in the earlier fault tree is now refined by an OR gate with the CCF event Nails as input. Thus, all tires fail when the event Nails occurs, causing the intermediate event Tires fail to occur. In spite of redundancy, this failure occurs as a result of a single CCF event.

The other input of each OR gate is a new event Tire1 indep, modeling the independent failure of the respective tire. For example, inputs of the gate Tire1 fails are Tire1 indep and Nails.

14.3   The β-Factor Model

The β-factor model [4] assumes that when a common cause occurs, all elements within the CCF group fail. This approach differentiates between independent failures and a single common cause failure that impacts the entire CCF group.

Updating the Fault Tree. This situation is modeled by updating the fault tree. For a CCF group $e_1, \ldots, e_n$, the original basic events are refined by an OR-gate with two new events as inputs. The first one models the common cause failure of all events from the CCF group. The second input models an independent failure of the original basic event, for example, due to wear. More precisely, each event e in the CCF group is refined through an OR-gate

$$e = \text{OR}(e^{\text{CCF}}, e^{\text{idp}})$$

These newly created events are called *CCF events*. Here, e^{CCF} represents the common cause failure and e^{idp} the independent failure of the component.

Quantifying Failure Probabilities. An important task is to derive probabilities of these new CCF events. That is, the failure probability p of the original event e should be split into a probability p^{CCF} of the CCF failure and the probability p^{idp} of the independent failure, such that $p_e = p^{\text{CCF}} + p^{\text{idp}}$.

The β-factor model determines the probability split by a single parameter $\beta \in [0, 1]$. It represents the percentage (or ratio) of the failure probability that is attributed to the common cause. The remaining fraction $1 - \beta$ is then the percentage of the independent failure:

$$p^{\text{CCF}} = \beta \cdot p$$
$$p^{\text{idp}} = (1 - \beta) \cdot p$$

Then, indeed,

$$p_e = \beta \cdot p_e + (1 - \beta) \cdot p_e = p^{\text{CCF}} + p^{\text{idp}}$$

Note that the factor β is the same for all events from this CCF group. With this treatment, standard methods for solving fault trees can be applied to the modified fault tree. Common cause failures then contribute to the analyzed dependability measure.

14.4 Common Cause Failures for Subsets of a CCF Group

In the β-factor model, a common cause event impacts the entire system with redundancies. Specifically, if such an event occurs, all dependent components are assumed to fail, regardless of the system's redundancy level. This effectively makes the CCF event a single point of failure in redundant systems: in the road trip example, Nails is a single point of failure for Tires fail.

Consequently, the β-factor model provides a conservative treatment of dependencies stemming from underlying failure mechanisms. However, it can be overly pessimistic by treating the common cause as a single point of failure within the CCF subgroup. Indeed, in a road trip scenario, not all tires would necessarily fail if there are nails on the road. To balance the need for reducing conservatism while avoiding the underestimation of actual risk, a different setup is proposed where common cause failures can also affect subsets of the CCF group. In the road trip example, either 1, 2, 3, 4, or all 5 tires may fail. These subsets are all assigned different probabilities. The α-factor and multiple Greek letter models provide common methods for splitting these probabilities.

Updating the Fault Tree. In general, for a CCF group $e_1, \ldots, e_n$, each original basic event e_i is replaced by an OR-gate with 2^{n-1} CCF events as inputs. That is,

$$e = \mathsf{OR}(S_1, S_2, \ldots, S_{2^{n-1}})$$

where all $S_1, S_2, \ldots, S_{2^{n-1}}$ represent (or, include) subsets of $\{e_1, \ldots, e_n\}$ that all contain e_i. There are 2^{n-1} of such subsets, hence the OR-gate has 2^{n-1} arguments. Each subset S models the simultaneous failure of e_i with other elements in S due to a common cause. For example, the set $\{e_1, e_3, e_7\}$ models the simultaneous failure of events e_1, e_3, and e_7. Note that this set appears as a child under e_1, e_3, and e_7. The subset $\{e_i\}$ represents the independent failure of e_i.

Example 41

Consider a CCF group consisting of three elements e_1, e_2, e_3. Then, CCF events are created as follows:

$$e_1 = \mathsf{OR}(\{e_1\}, \{e_1, e_2\}, \{e_1, e_3\}, \{e_1, e_2, e_3\})$$
$$e_2 = \mathsf{OR}(\{e_2\}, \{e_1, e_2\}, \{e_2, e_3\}, \{e_1, e_2, e_3\})$$
$$e_3 = \mathsf{OR}(\{e_3\}, \{e_1, e_3\}, \{e_2, e_3\}, \{e_1, e_2, e_3\})$$

Figure 14.1 shows an update of a fault tree with CCF events. Note that each subset of $\{e_1, e_2, e_3\}$ represents a CCF event which acts as a basic event in the updated fault tree and that e_i itself has become an intermediate event.

14.4.1 Quantifying the CCF Events

Again, we need to quantify the CCF events newly created from a CCF group with n basic events $e_1, \ldots, e_n$, redistributing the failure probability Q of the original basic event e_i over the subsets $S_1, S_2 \ldots, S_m$, where $m = 2^{n-1}$, that contain the event e_i. We present two parametric models to do so: the α-factor model and the multiple Greek letter model. Both models use the number of elements in the sets S_i, and assume that all original basic events e_i have the same failure probability.

The α-Factor Model. The α-factor model [5] uses parameters $\alpha_1, \ldots, \alpha_n$ for a CCF group with n basic events. Further, it is assumed that all basic events from the CCF group have the same failure probability Q. Then the probability of a CCF event S_i with k basic events is given by:

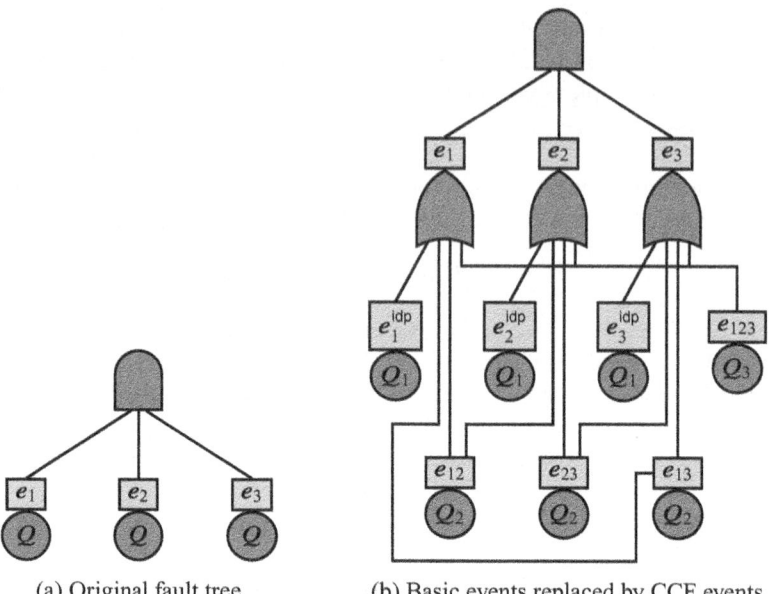

(a) Original fault tree   (b) Basic events replaced by CCF events

Fig. 14.1 Illustration of a fault tree modification using all subsets of the CCF group

$$Q_k = \frac{k}{\binom{n-1}{k-1}} \cdot \frac{\alpha_k}{\alpha_{\text{Total}}} \cdot Q$$

$$\alpha_{\text{Total}} = \sum_{i=1}^{n} i \cdot \alpha_i$$

It can then be shown that the probabilities of the CCF events $S_1, S_2 \ldots, S_m$ constitute a partitioning of the original probability: $Q = \sum_{k=1}^{n} \binom{n-1}{k-1} \cdot Q_k$. Note that there are $\binom{n-1}{k-1}$ subsets with k basic events which include one specific basic event e_i.

The Multiple Greek Letter Model. The multiple Greek letter model [6] extends the β model by complementing β with additional parameters for CCF events with different numbers of included basic events. Parameters of this model are often denoted by $\beta, \gamma, \delta, \ldots$. For a CCF group with n basic events, we denote them by $\rho_1 = 1, \rho_2 = \beta, \rho_3 = \gamma, \ldots, \rho_{n+1} = 0$. We again assume that all basic events from the CCF group have the same failure probability Q. The probability of a CCF event S_i with k basic events ($1 \leq k \leq n$) is given by the formula:

$$Q_k = \frac{1}{\binom{n-1}{k-1}} \cdot \left(\prod_{i=1}^{k} \rho_i \right) \cdot (1 - \rho_{k+1}) \cdot Q$$

Example 42

Let us consider a CCF group with three basic events e_1, e_2, e_3 and with parameters $\alpha_1 = 0.79$, $\alpha_2 = 0.2$, $\alpha_3 = 0.01$. Each e_i has the failure probability Q. Then, we obtain

$$Q_1 = 0.648 \cdot Q$$
$$Q_2 = 0.164 \cdot Q$$
$$Q_3 = 0.025 \cdot Q$$

The logical structure of the OR-gates with CCF events that replace the original basic events is the same as in Fig. 14.1. For the event e_1, there is one set of cardinality one that includes e_1, namely $\{e_1\}$, two sets of cardinality two that include e_1, namely $\{e_1, e_2\}$ and $\{e_1, e_3\}$, and one set of cardinality three that contains e_1, namely $\{e_1, e_2, e_3\}$. Summing up probabilities of these sets gives the original failure probability Q.

$$Q_1 + 2 \cdot Q_2 + Q_3 = (0.648 + 2 \cdot 0.164 + 0.025) \cdot Q = Q$$

References

1. Mosleh A (1998) Procedures and guidelines in modeling common cause failures in probabilistic risk assessment. NUREG/CR-5485
2. International Atomic Energy Agency (1992) Procedures for Conducting Common Cause Failure Analysis in Probabilistic Safety Assessment. No. 648 in TECDOC Series, IAEA, Vienna
3. Mosleh A, Fleming KN, Parry GW, Paula HM, Worledge DH, Rasmuson DM (1988) Procedures for treating common cause failures in safety and reliability studies: Volume 1, procedural framework and examples: Final report. NUREG/CR-4780
4. Fleming KN (1974) Reliability model for common mode failures in redundant safety systems. In: General atomic report GA-A13284
5. Mosleh A, Siu N (1987) A multi-parameter common cause failure model. In: Proceedings of the 9th International Conference on Structural Mechanics in Reactor Technology
6. Fleming KN, Kalinowski AM (1983) An extension of the beta factor method to systems with high levels of redundancy. In: Pickard, Lowe and Garrick, Inc., PLG-0289

Non-coherent Fault Trees

15

Fault trees considered thus far, featuring AND-, OR- and VOTING-gates are *coherent*. This means that, whenever the system has failed, additional occurrences of any basic event will never result in a state where the system resumes functioning. In terms of cut sets: if we add more basic events to a cut set, the new set remains a cut set. This coherence property might not hold if we use negated gates, which include the NOT-gate and NOT-based gates such as the NAND, NOR, and XOR-gate, see Fig. 15.1.

From a purely logical perspective, non-coherency implies that a system functions better if additional failures occur. This seems counter-intuitive and, indeed, systems are typically not designed this way. Nevertheless, negated gates do have their place in fault trees in order to exclude event combinations that we wish to ignore in our analysis. Typical purposes include the modeling and analysis of maintenance plans, mutually exclusive operational modes, and specific failure scenarios, especially coming from event trees.

This chapter describes the practical usage of negated gates for various purposes, as well as their analysis methods. Non-coherent fault trees require additional care and clarity about the role of not-logic in the modeling, especially when it comes to minimal cut sets. An important difference to coherent fault trees is that the list of minimal cut sets no longer characterize the fault tree logic.

© The Author(s), under exclusive license to Springer Nature Switzerland AG 2026   191
M. Stoelinga et al., *Concise Guide to Fault Tree Analysis*, Computer Science Foundations and Applied Logic, https://doi.org/10.1007/978-3-031-78287-9_15

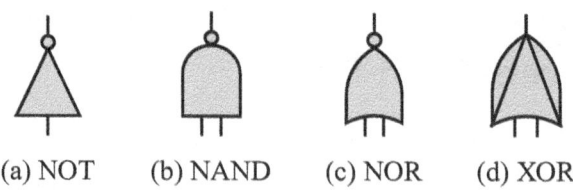

(a) NOT   (b) NAND   (c) NOR   (d) XOR

Fig. 15.1 Negated gates used in non-coherent fault trees: NOT, negated AND (NAND), negated OR (NOR), and exclusive OR (XOR)

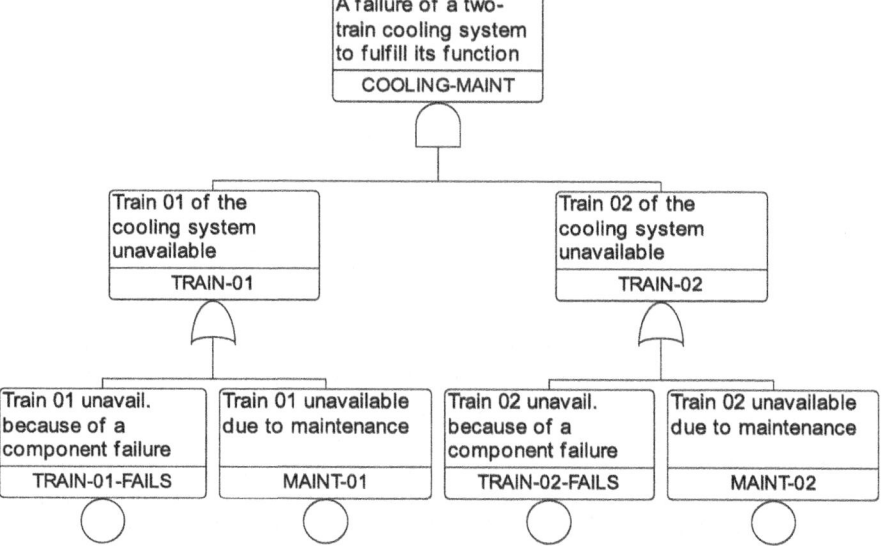

Fig. 15.2 Coherent fault tree F_{org} with two redundant subsystems that can be unavailable due to maintenance

15.1 Modeling with Not-Logic

We start by illustrating the use of not-logic for one of its common purposes, namely the modeling of maintenance aspects.

15.1.1 Maintenance

NOT-gates often occur in the modeling of maintenance. We illustrate its application using the cooling system introduced in Chap. 6, where each train can be unavailable due to maintenance. In this section, we simplify the example to two trains, and treat their failures due to maintenance as basic events, as depicted in Fig. 15.2. Their probabilities express the fraction of time in which the respective train is under maintenance.

This system has four minimal cut sets:

$$C_1 = \{\text{TRAIN-01-FAILS, TRAIN-02-FAILS}\}$$
$$C_2 = \{\text{MAINT-01, TRAIN-02-FAILS}\}$$
$$C_3 = \{\text{TRAIN-01-FAILS, MAINT-02}\}$$
$$C_4 = \{\text{MAINT-01, MAINT-02}\}$$

However, the minimal cut set $C_4 = \{\text{MAINT-01, MAINT-02}\}$ represents a failure scenario in which both trains are unavailable due to maintenance at the same time. This is typically never the case: safety relevant technical systems have a well-designed maintenance plan that avoids simultaneous maintenance of both trains.

Since maintenance happens frequently, especially compared to component failures, this cut set will have a high failure probability, which will often dominate the minimal cut set list. Since the set $\{\text{MAINT-01, MAINT-02}\}$ does not represent a valid failure scenario, it should not be included in any minimal cut set. We present two solutions to exclude C_4: restructuring the fault tree while keeping it coherent and using negated logic, yielding a non-coherent tree.

Solution 1: Restructuring via Coherent Fault Tree. One way to exclude the minimal cut set $\{\text{MAINT-01, MAINT-02}\}$ is by modifying the fault tree, as shown in Fig. 15.3. This fault tree excludes $C_4 = \{\text{MAINT-01, MAINT-02}\}$ as a minimal cut set, while keeping C_1, C_2, and C_3.

Any minimal cut set C can be excluded in this way, i.e., it is always possible to propose a coherent fault tree that has exactly the same minimal cut sets, except for C. In the ultimate case, one rewrites the fault tree in its disjunctive normal form (DNF), as elaborated in Sect. 7.3—note that F_{mod} is in DNF. A disadvantage of this solution is that the structure of the modified fault tree might no longer correspond to the structure of the system under analysis, making the modified tree more difficult to understand.

Furthermore, removing the set $\{\text{MAINT-01, MAINT-02}\}$ from the list of *minimal* cut sets does not remove all combinations that contain these two events from the list of cut sets. For instance, the set $\{\text{TRAIN-01-FAILS, MAINT-01, MAINT-02}\}$ is still a cut set in F_{mod}, albeit not a minimal one.

Since F_{mod} and F_{org} are coherent fault trees, and the minimal cut sets of F_{mod} are included in the minimal cut sets of F_{org}, we have that

$$\mathbb{P}[F_{\text{mod}}] \leq \mathbb{P}[F_{\text{org}}]$$

We argue that $\mathbb{P}[F_{\text{mod}}]$ is a more realistic failure probability than $\mathbb{P}[F_{\text{org}}]$, since $\mathbb{P}[F_{\text{mod}}]$ excludes the high-probability minimal cut set $\{\text{MAINT-01, MAINT-02}\}$. In other words, the probability $\mathbb{P}[F_{\text{org}}]$ is too conservative. As usual, $\mathbb{P}[F_{\text{mod}}]$ can be approximated via its rare event approximation (i.e., sum of minimal cut set probabilities) or the min cut upper bound.

Solution 2: Fault Tree with Negations. An alternative modeling approach to exclude C_4 uses negation, shown in Fig. 15.4. The non-coherent tree F_{nc} extends F_{org} by an

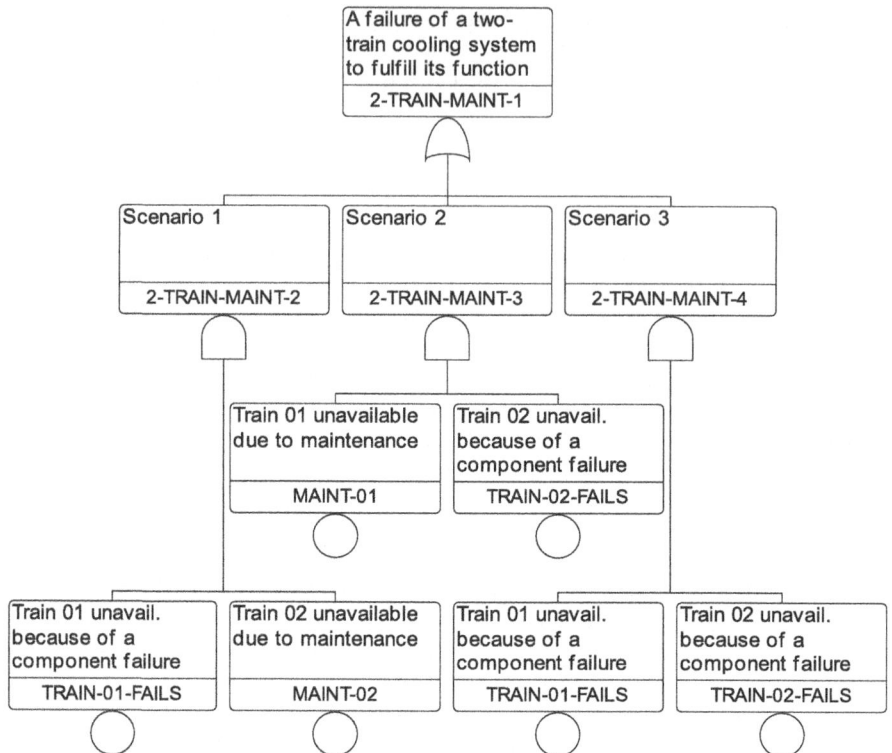

Fig. 15.3 Coherent fault tree F_{mod} with two redundant subsystems that can be unavailable due to maintenance, where simultaneous maintenance is excluded as a failure scenario

intermediate event MAINT-EXCL expressing that both trains cannot be maintained simultaneously.

The minimal cut sets generated by F_{nc} are the same as the minimal cut sets generated by F_{mod}, namely C_1, C_2, C_3. However, unlike F_{mod}, the F_{nc} does not include {MAINT-01, TRAIN-02-FAILS, MAINT-02} as a cut set. This shows that, for non-coherent fault trees, the minimal cut sets do not tell the entire story. Moreover, a (minimal) cut set is no longer a set that makes the fault tree fail, since some basic events may be required to not fail.

Implicants take the role of cut sets, being a sufficient condition for the fault tree to fail. Prime implicants are the smallest implicants. The following are prime implicants of F_{nc}, written in logical notation:

$$\text{MAINT-01} \land \text{TRAIN-02-FAILS} \land \neg\text{MAINT-02}$$

$$\text{TRAIN-01-FAILS} \land \text{TRAIN-02-FAILS}$$

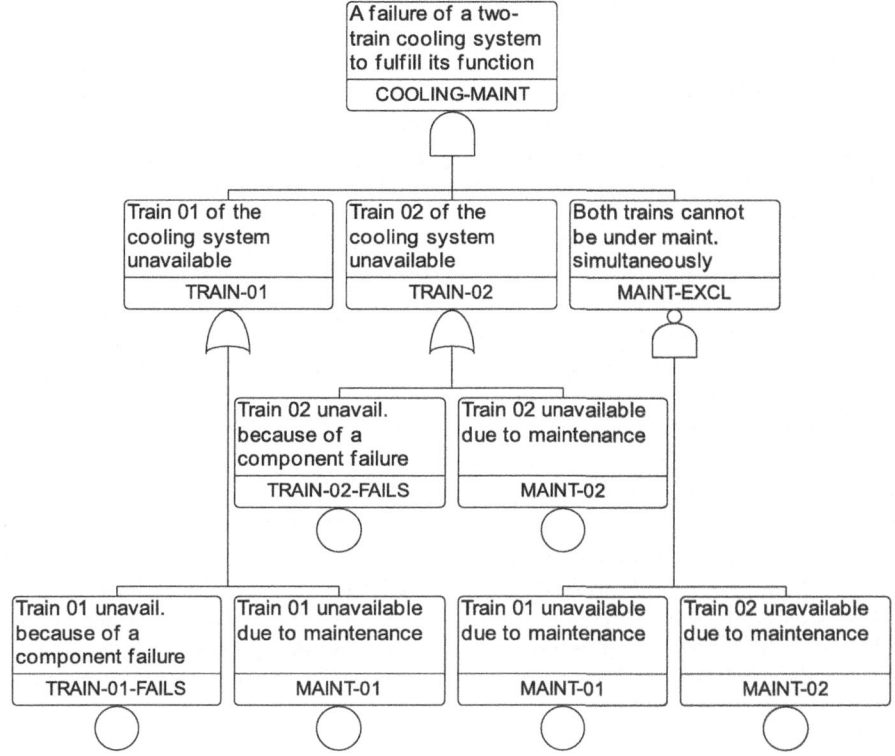

Fig. 15.4 Non-coherent fault tree F_{nc} with two redundant subsystems that excludes simultaneous maintenance of both trains by a negated gate

The following is an implicant, but not prime:

$$\text{TRAIN-01-FAILS} \wedge \text{TRAIN-02-FAILS} \wedge \neg\text{MAINT-01}$$

Since F_{nc} has fewer cut sets than F_{mod}, it has a lower top probability than F_{mod}, which has a lower top probability than F_{org}, as we saw before.

$$\mathbb{P}[F_{nc}] \leq \mathbb{P}[F_{mod}] \leq \mathbb{P}[F_{org}]$$

15.1.2 Purposes for Not-Logic

Several scenarios exist where not-logic is useful, restricting relevant failure scenarios and making analysis more precise.

Maintenance. Operational instructions may state that two redundant components shall never be maintained simultaneously. Therefore, we can screen out the probability of a simultaneous unavailability of both components due to maintenance. Negations might provide a concise way to express this.

Mutually Exclusive Operational Scenarios. Basic events may model failure scenarios that are mutually exclusive. Typical examples are a full power state versus a shut down state; summer versus winter conditions; a pump being on or off. These states are mutually exclusive and can be modeled via a XOR-gate.

If-Then-Else Operational Scenarios. A system's failure behavior can vary depending on the success or failure of another interconnected system. Consider a scenario involving two subsystems within a larger system: one provides power, while the other serves as a standby passive cooling system.

For the passive cooling system to activate, a valve must first open. If the power supply fails, the maneuvering power for this valve also fails. However, the valve is equipped with a spring mechanism that automatically opens it in the event of a power failure. In contrast, a scenario without power necessitates manual intervention for a different system. If the maneuvering power supply is functional, the valve introduces an additional failure mode—failure to open—because the valve mechanism might still fail despite the available power supply.

The situation can be expressed logically as follows: If the power supply fails, the passive cooling system will start automatically; otherwise, the system may fail to start if the valve fails to open. The logical expression for this scenario is (POWER-FAILS $\land$ MANUAL-ACTION) $\lor$ ($\neg$POWER-FAILS $\land$ FAILURE-TO-OPEN).

Encoding of Accident Scenarios. A description of an accident scenario might also include information about success of certain safety systems. This can also be relevant for the analysis. For example, modeling success of systems with relatively high failure probabilities caused by seismic events or physical phenomena allows an analysis tool to assess the success probability also quantitatively. Moreover, different safety systems might be required depending on success or failure of the first-line safety system.

15.2 The Behavior of Non-coherent Fault Trees

Table 15.1 presents the behavior for non-coherent gates. The logical behavior defines for each gate its output b_O as a function of its inputs $b_1, \ldots, b_n$. Unlike coherent fault trees, non-coherent fault trees can encode any Boolean function.

The probabilistic behavior defines the output probability p_O as a function of the input probabilities $p_1, \ldots, p_n$, provided that inputs are stochastically independent.

Table 15.1 Logical and probabilistic behavior in non-coherent fault trees. Here, b_i denotes the status of event E_i, and p_i the probability of E_i

Symbol	Name	Probability
	NOT-gate	The output occurs if its input does not occur $f_O(b_1) = \neg b_1$ $p_O = 1 - p_1$
	NAND-gate	Output occurs if at least one of the inputs does not occur $f_O(b_1, \ldots, b_n) = \neg(b_1 \vee \ldots \vee b_n) = \neg b_1 \wedge \ldots \wedge \neg b_n$ $p_O = 1 - (p_1 \cdot p_2 \cdots p_n)$
	NOR-gate	O occurs if none of the b_i occurs $f_O(b_1, \ldots, b_n) = \neg(b_1 \wedge \ldots \wedge b_n) = \neg b_1 \vee \ldots \vee \neg b_n$ $p_O = (1 - p_1)(1 - p_2)\cdots(1 - p_n)$
	XOR-gate	O occurs if exactly one of its inputs occurs For two inputs $f_O(b_1, b_2) = (b_1 \wedge \neg b_2) \vee (\neg b_1 \wedge b_2)$ $p_O = p_1(1 - p_2) + p_2(1 - p_1)$

15.2.1   Structure Function

Non-coherent fault trees still represent Boolean formulas and can thus be expressed through their structure function Φ_F. As before, $\Phi_F(\mathbf{b}, E)$ indicates whether event E fails given the status vector $\mathbf{b}$. We derive the structure function Φ_F for a non-coherent fault tree F by interpreting F as a logical formula, adhering to the gate rules outlined in Table 15.1, in addition to the rules for coherent fault trees outlined in Table 10.1.

Definition 24 Let F be a fault tree, $\mathbf{b} = (b_1, b_2, \ldots, b_n)$ a status vector. For a non-basic event E, we write $Gate(E)$ for the gate type of event E and $E_1, E_2, \ldots, E_m$ for its children. Then, the structure function $\Phi_F : \{0, 1\}^n \times V \rightarrow \{0, 1\}$ of F is defined as follows:

$\Phi_F(\mathbf{b}, E) =$

$$\begin{cases}
\Phi_F(\mathbf{b}, E) = b_i & \text{if } E = e_i \in BE \\
\Phi_F(\mathbf{b}, E) = \Phi_F(\mathbf{b}, E_1) \wedge \Phi_F(\mathbf{b}, E_2) \wedge \cdots \Phi_F(\mathbf{b}, E_m) & \text{if } Gate(E) = \text{AND} \\
\Phi_F(\mathbf{b}, E) = \Phi_F(\mathbf{b}, E_1) \vee \Phi_F(\mathbf{b}, E_2) \vee \cdots \Phi_F(\mathbf{b}, E_m) & \text{if } Gate(E) = \text{OR} \\
\Phi_F(\mathbf{b}, E) = (\sum_{i=1}^{m} \Phi_F(\mathbf{b}, E_i)) \geq k & \text{if } Gate(E) = \text{VOT}(k/N) \\
\Phi_F(\mathbf{b}, E) = \neg(\Phi_F(\mathbf{b}, E_1) \wedge \Phi_F(\mathbf{b}, E_2) \wedge \cdots \Phi_F(\mathbf{b}, E_m)) & \text{if } Gate(E) = \text{NAND} \\
\Phi_F(\mathbf{b}, E) = \neg(\Phi_F(\mathbf{b}, E_1) \vee \Phi_F(\mathbf{b}, E_2) \vee \cdots \Phi_F(\mathbf{b}, E_m)) & \text{if } Gate(E) = \text{NOR}
\end{cases}$$

As before, we write $\Phi_F(\mathbf{b}) = \Phi_F(\mathbf{b}, Top)$.

The structure function can still be converted into a BDD by Shannon expansion:

$$\Phi_F(b_1, b_2, \cdots, b_n) = (\neg b_1 \wedge \Phi_F(0, b_2, \cdots, b_n)) \vee (b_1 \wedge \Phi_F(1, b_2, \cdots, b_n))$$

15.2.2 Probabilistic Behavior

The probabilistic behavior of a non-coherent fault tree is still obtained via the status vector probabilities. That is, the probability for event E is obtained by summing all status vector probabilities that make event E fail, i.e., all vectors $\mathbf{b}$ with $\Phi_F(\mathbf{b}, E) = 1$.

Definition 25 Let $\mathbf{b} = (b_1, b_2, \ldots, b_n)$ be a status vector and $\mathbf{p} = (p_1, p_2, \ldots, p_n)$ be a vector of basic event probabilities. The probability of $\mathbf{b}$, denoted $\mathbb{P}[\mathbf{b}]$, is given by

$$\mathbb{P}[\mathbf{b}] = \prod_{1 \leq i \leq n} b_i \cdot p_i + (1 - b_i) \cdot (1 - p_i)$$

$$\mathbb{P}[E] = \mathbb{P}[\Phi_F(\mathbf{b}, E) = 1] = \sum_{\mathbf{b} \in \{0,1\}^n} \mathbb{P}_\mathbf{p}[\mathbf{b}] \cdot \Phi_E(\mathbf{b}, E)$$

The following table gives the probability for the fault tree $F = \text{XOR}(b_1, b_2)$.

b_1	b_2	$\text{XOR}(b_1, b_2)$	probability
0	0	0	
0	1	1	$\overline{p_1} p_2$
1	0	1	$p_1 \overline{p_2}$
1	1	0	
			$\overline{p_1} p_2 + p_1 \overline{p_2}$

Computing Top Probabilities. Top event probabilities for non-coherent fault trees are calculated using the same methods as for coherent fault trees. For tree-shaped fault trees, the event probabilities can still be analyzed via the bottom-up algorithm, using the probabilistic rules in Table 15.1. For DAG-shaped fault trees, the top event probability can be computed via BDD-based methods. The structure function for non-coherent fault trees remains a Boolean function, it can be converted into a Binary Decision Diagram (BDD) for this purpose.

15.3 Cut Sets and Implicants

The most crucial difference between coherent and non-coherent fault trees is that cut sets work differently in non-coherent fault trees. Conceptually, the definition of a (minimal) cut set in a non-coherent fault tree remains the same. A cut set is a set C of basic events such that $\Phi_F(C) = 1$, and a minimal cut set is a cut set where no elements can be removed.[1] However, there are three key differences between coherent and non-coherent fault trees regarding cut sets.

- Adding additional elements to a cut set may yield a set that is no longer a cut set.
- If all elements in a cut set fail, it is no longer guaranteed that the fault tree fails.
- A fault tree is no longer characterized by its minimal cut sets, i.e., its structure function cannot be reconstructed from the minimal cut set list.

We illustrate these points via $F = \mathrm{XOR}(e_1, e_2)$. This tree has two minimal cut sets $C_1 = \{e_1\}$ and $C_2 = \{e_2\}$, which are the only two cut sets.

- Adding the event e_2 to cut set $C_1 = \{e_1\}$ makes C_1 lose its status as a cut set.
- The set $\{e_1, e_2\}$ fails all elements in C_1, but this set does not fail the fault tree itself.
- The fault trees $\mathrm{XOR}(e_1, e_2)$ and $\mathrm{OR}(e_1, e_2)$ have exactly the same minimal cut sets, but their diagrams and structure functions are different. Thus, knowing all minimal cut sets does not allow one to reconstruct the fault tree logic.

Implicants and Prime Implicants. The key issue is that in non-coherent fault trees, one must not only know which events must fail, but also which events have not failed. The latter is included in the notion of *implicant*.

Definition 26 An *implicant* consists of two sets of basic events $C = \{e_1, \ldots, e_n\} \subseteq BE$ and $D = \{e'_1, \ldots, e'_m\} \subseteq BE$, such that if all events in C fail, and all events in D

---

[1] Recall that the notation $\Phi_F(C) = 1$ is shorthand for the status vector **b**, where $b_i = 1$ if $b_i \in C$ and $b_i = 0$ if $b_i \notin C$, i.e., $\Phi_F(C) = 1$ assumes that elements not in C are operational.

do not fail, then the fault tree fails.

$$\forall i \le n, j \le m \quad b_{e_i} = 1, b_{e_j} = 0 \Rightarrow \Phi_F(\mathbf{b}) = 1$$

Here b_{e_i} denotes the status bit corresponding to basic event e_i. A *prime implicant* is an implicant that cannot be reduced, that is, no elements can be omitted from either C or D.

Note that if C, D is implicant, then C is a cut set; and if C, D is a prime implicant, then C is a minimal cut set. Prime implicants have similar properties as minimal cut sets in coherent fault trees:

- By adding new events to an implicant, the pair of sets remains an implicant— assuming that one does not add elements of C to D or vice versa.
- If all events in C fail and events in D do not fail, then the fault tree is guaranteed to fail, irrespective of the status of events not contained in C or D.
- The fault tree is fully characterized by its prime implicants, that is, the structure function can be reconstructed from a list of prime implicants.

In the literature, it is common to encode implicants as Boolean formulas. Then, an implicant is a conjunction of basic events and negated basic events that implies the structure function. A prime implicant is the smallest implicant. That is, an implicant is a Boolean formula $e_1 \wedge e_2 \wedge \ldots e_n \wedge \neg e'_1 \wedge \neg e'_2 \wedge \ldots \neg e'_n$ such that

$$e_1 \wedge e_2 \wedge \ldots e_n \wedge \neg e'_1 \wedge \neg e'_2 \wedge \ldots \neg e'_n \Rightarrow \Phi_F^{\text{Bool}}$$

Here, Φ_F^{Bool} is the structure function encoded as a Boolean expression. A prime implicant is an implicant that has no subformula that is an implicant.

Example 43

For example, the fault tree $F = \text{OR}(\text{XOR}(E_1, E_2), \text{OR}(E_3, E_4))$ has four minimal cut sets

$$\{E_1\}, \{E_2\}, \{E_3\} \text{ and } \{E_4\}$$

However, these cut sets are of different nature: For E_1 to cause a top failure, E_2 must not fail, and this information is not present in the cut set $\{E_1\}$. Therefore, cut sets $\{E_1\}$ and $\{E_2\}$ are less likely to induce a top event failure than $\{E_3\}$ and $\{E_4\}$.

Then F has four prime implicants:

$$e_1 \wedge \neg e_2, \quad \neg e_1 \wedge e_2, \quad e_3, \text{ and } \quad e_4$$

The expression $e_1 \wedge e_4 \wedge \neg e_2$ is an implicant, but not prime, and so is $e_1 \wedge \neg e_4 \wedge \neg e_2$.

15.3.1 Probabilities of Implicants

Let $\mathbb{P}[e]$ denote the probability of basic event e. Then, the probability of an implicant $\langle C, D \rangle$ is given by

$$\mathbb{P}[\langle C, D \rangle] = \prod_{e \in C} \mathbb{P}[e] \prod_{e \in D} (1 - \mathbb{P}[e])$$

Approximation of Top Probability. The prime implicant probabilities can be used to approximate the top probability of a non-coherent fault tree, in the same way as we use the cut set probabilities in a coherent fault tree. That is, we can generalize the rare event approximation (REA) to prime implicants.

$$\mathbb{P}[Top] \leq \sum_{\langle C, D \rangle \in \mathcal{P}} \mathbb{P}[\langle C, D \rangle] \qquad \text{(REA)}$$

Here, $\mathcal{P}$ is the set of all prime implicants. However, the *MCUB-* approximation does not hold for prime implicants. That is, there are cases where

$$\mathbb{P}[Top] \not\leq 1 - \prod_{\langle C, D \rangle \in \mathcal{P}} (1 - \mathbb{P}[C]) \mathbb{P}[D]$$

For example, the non-coherent fault tree $F = \text{XOR}(e_1, e_2)$. This fault tree has two prime implicants, namely $\langle \{e_1\}, \{e_2\} \rangle$ and $\langle \{e_2\}, \{e_1\} \rangle$. If we take $\mathbb{P}[e_1] = \mathbb{P}[e_2] = \frac{1}{2}$, then we have:

$$\mathbb{P}[Top] = \mathbb{P}[e_1](1 - \mathbb{P}[e_2]) + (1 - \mathbb{P}[e_1])\mathbb{P}[e_2] = 1/2 \cdot 1/2 + 1/2 \cdot 1/2 = 1/2$$

$$MCUB(F) = 1 - (1 - \mathbb{P}[e_1](1 - \mathbb{P}[e_2]))(1 - (1 - \mathbb{P}[e_1])\mathbb{P}[e_2])$$

$$= 1 - (1 - \frac{1}{2} \cdot \frac{1}{2})(1 - \frac{1}{2} \cdot \frac{1}{2})$$

$$= 1 - \frac{3}{4} \cdot \frac{3}{4}$$

$$= \frac{7}{16}$$

$$< \mathbb{P}[Top]$$

15.4 Minimal Cut Set-Based Analysis of Not-Logic

Top event probabilities for non-coherent fault trees are calculated or approximated using methods similar to those for coherent fault trees. For tree-shaped fault trees, probabilities can still be analyzed using a bottom-up approach with probabilistic rules. For DAG-shaped fault trees, BDD-based methods are used by converting the structure function into a Binary Decision Diagram (BDD). The algorithm remains consistent with the one in Chap. 9.

Another approach is to use prime implicants or minimal cut sets. In practice, minimal cut set probabilities are often preferred for non-coherent fault trees, as they provide a useful approximation, particularly when success probabilities are high and have little impact on the result. Minimal cut sets offer a close coherent over-approximation of satisfying assignments [1] and effectively describe failure scenarios, making them sufficient for qualitative analysis.

However, generating minimal cut sets for non-coherent fault trees requires adjustments to handle negated gates correctly. The MOCUS algorithm does not address how to expand negated gates during cut set generation, and applying the complement operation in the MICSUP algorithm would be computationally intensive. BDD approaches still work for non-coherent fault trees.

15.4.1   Generating Minimal Cut Sets

As discussed in Sect. 15.1, not-logic provides a compact method for excluding specific combinations of failures. Essentially, the failure logic of the system, which encompasses all possible failure combinations, is captured within the coherent part of the fault tree. Negated gates are used to exclude certain combinations from consideration. An algorithm designed to generate minimal cut sets can therefore concentrate on the coherent part of the fault tree, using the entire fault tree to validate the minimal cut sets produced.

Validation. Each minimal cut set generated from the coherent part of the fault tree undergoes a validation process against the original fault tree. This involves checking whether a minimal cut set derived from the coherent part also qualifies as a cut set in the original fault tree. Validation is performed by propagating the cut set through the original fault tree, where all basic events not included in the cut set are set to False. If the top gate evaluates to True, the minimal cut set is valid for the original fault tree. If not, it violates the not-logic constraints and is discarded.

This algorithm has the following properties:

- *Minimality and Validity*. The algorithm ensures that the generated cut sets are minimal and valid for the original fault tree.
- *Completeness for the Coherent Part*. The algorithm generates all minimal cut sets from the coherent part that also cause the top event to fail in the original fault tree.

15.4.2   Quantifying Success

As mentioned earlier, success probabilities are often very high and may not significantly affect approximation results. However, this is not always the case, especially in safety-critical and highly reliable systems. Sometimes, modeling system configurations or operating modes probabilistically is necessary. For example, in if-then-else scenarios, where one configuration is specified and the rest are negated, the success probability can significantly impact the top value. If a system is in the configuration of interest 30% of the time, the "success" probability (i.e., the system being in a different configuration) would be 0.7.

In other cases, such as common cause initiators or external hazards like seismic events, fire, or flooding, success probabilities can also be substantial. For example, component failures following an earthquake might have high probabilities depending on the seismic event's peak ground acceleration.

In cut set-based solutions, new basic events (or modules [2]) can be introduced to represent the probability of success for selected negated gates, often arising from event tree modeling. The success probability is conservatively approximated by taking the failure probability of the negated gate and complementing it numerically: $\mathbb{P}[Success] = 1 - \mathbb{P}[Failure]$.

In event tree models, success can also be estimated by calculating failure probabilities for all internal nodes and complementing the failure branch to quantify the success branch from the same node [3].

References

1. Rauzy A (2001) Mathematical foundations of minimal cutsets. IEEE Trans Reliab 50(4):389–396
2. RiskSpectrum AB (2024) RiskSpectrum analysis tools theory manual, version 4.0.0
3. Nusbaumer O, Rauzy A (2013) Fault tree linking versus event tree linking approaches: a reasoned comparison. Proc. Inst. Mech. Eng., Part O: J. Risk Reliabil. 227(3):315–326

Beyond Static Fault Trees

16

This book has presented standard fault trees, which are widely used in practical, industrial risk management. While these are powerful enough for many practical applications, various extensions have been developed to support more detailed modeling than the Boolean model allows.

This chapter considers some models beyond standard fault trees, broadly divided into four categories:

- Section 16.1 looks at extensions supporting more complex combinations between events than just the Boolean gates described earlier in this book.
- Section 16.2 treats extensions that (mostly) use Boolean gates, but support more complex behavior in basic events.
- Some techniques have been developed to apply fault trees effectively in specific domains and situations, which are described in Sect. 16.3.
- Section 16.4 briefly describes some models from risk management that are not fault trees, but that are closely related.

16.1 Extended Relation Between Events

16.1.1 Dynamic Fault Trees

One of the most notable expansions is the dynamic fault tree model, which integrates time-dependent behavior. All gates discussed thus far are static, meaning that only the occurrence of failure events matters, not their temporal order. These static (or standard) fault trees are favored for their simplicity, enabling the analysis of very large fault trees while encompassing a broad spectrum of phenomena.

© The Author(s), under exclusive license to Springer Nature Switzerland AG 2026   205
M. Stoelinga et al., *Concise Guide to Fault Tree Analysis*, Computer Science Foundations
and Applied Logic, https://doi.org/10.1007/978-3-031-78287-9_16

Fig. 16.1 Gates in dynamic
fault trees

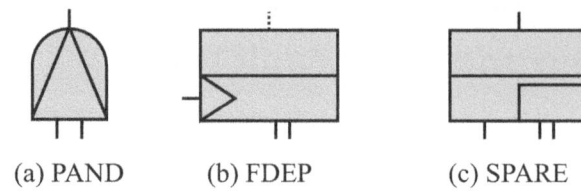

(a) PAND   (b) FDEP   (c) SPARE

However, static fault trees lack the capability to represent dynamic behavior and intricate interactions, which can be pivotal in various dependability scenarios, such as spare management and functional dependencies. Dynamic fault trees address such time dependencies by introducing new types of gates. Given their dynamic nature, analyzing dynamic fault trees necessitates more intricate analytical procedures.

Dynamic fault trees [1] extend standard fault trees by adding dynamic gates. The coherent and non-coherent gates in Table 2.1 and Fig. 2.5 are static: It only matters whether a gate has failed or not. In dynamic gates, also the failure order matters. Three dynamic gate types are common, see Fig. 16.1: The PAND fails if the inputs fail from left to right. The FDEP-gate consists of a trigger and several dependent events. Whenever the trigger fails, all dependent events fail too. For example, the outbreak of a fire makes all equipment in one building fail; if a power cable fails, all connected elements cannot perform their function either. The SPARE-gate models the behavior of spare components, and caters especially for the fact that spare components fail less frequently when they are not in use. Only after their primary fails, they become active and switch to a higher failure rate. Thus, the SPARE-gate features one primary and several spare inputs. The primary input is active at the start, while the spare inputs are passive (a.k.a. dormant). Whenever the primary fails, the first available spare becomes active. When no spares are available, then the SPARE-gate fails.

Given their dynamic nature, the analysis of dynamic fault trees requires very different techniques to handle the time dependencies. These techniques are often based on continuous time Markov chains and their variants. We refer the reader to [2,3] for details. A special class of these methods, e.g., [4–7] focuses on subsets of dynamic fault trees where the analysis first translates the dynamic fault tree into a static one, then solves this static fault tree by a minimal cut set decomposition, and finally quantifies minimal cut sets dynamically. This approach brings scalability of static fault tree analysis as an advantage in exchange for restricting the expressive power of dynamic fault trees and possibly applying further approximations.

Pandora Temporal Fault Trees An alternative approach to capture dynamic system behavior in a fault tree is the Pandora formalism [8]. This approach extends fault trees with Priority-AND, Simultaneous-AND, and Priority-OR gates. This allows description of how different failure sequences of the same basic events can have different effects on system failure.

A benefit of the Pandora method is that analysis can be performed by generalizing (minimal) cut sets to "(minimal) cut sequences," which are like cut sets except they can include elements of the form "$A < B$" denoting that event A must happen before event B in order for the cut sequence to occur. This analysis method allows

Fig. 16.2 Attack tree
modeling a bank robbery. To
rob a bank, attackers must
break in, open the safe, and
escape (in that order). The
safe is opened by cutting it
open, or by unlocking via
obtaining the key and
combination

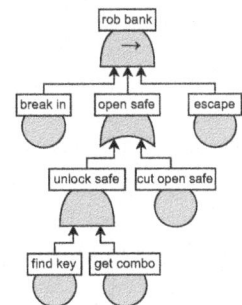

practical analysis of large fault trees, for which most analysis techniques for dynamic
fault trees would be computationally infeasible. The expressive power of Pandora
is narrower than that of dynamic fault trees, as the SPARE-gate does not have an
analog in Pandora.

16.1.2   Attack Trees and Attack-Fault Trees

Attack trees [9, 10] were invented as the equivalent of fault trees for cybersecurity,
representing how vulnerabilities can be exploited and combined into attack scenarios,
see Fig. 16.2 for an example. Like fault trees, an attack tree starts with a top event,
representing the attacker's goal. The top event can be refined into subgoals via gates
and the leaves of an attack tree are called basic attack steps.

  In their very basic form, attack trees and fault trees are very similar, both being
AND/OR trees. Despite their similarity to fault trees, attack trees feature several
remarkable subtrees [11]:

Different Gates: AND. The AND-gate in attack trees is often a sequential AND-gate
(SAND-gate). That is, the children of the AND-gate must be executed from left to
right. The top gate in Fig. 16.2 is a SAND-gate, indicated by the arrow inside gate.
The attacker must first break in, then open the safe, and finally escape; in that order.
The subgoal *Unlock safe* is a regular AND-gate, since unlocking the safe requires
both a key and a combo that can be executed in any order. For some analysis, the
difference between AND and SAND does not matter.

  (Note that the SAND-gate is also different from the PAND-gate in dynamic fault
trees: the SAND-gate executes the children from left to right. In the PAND-gate, all
children happen at the same time, but the failure only propagates if they fail in the
right order.)

Different Gates: OR. The OR-gate in attack trees also behaves differently from the
OR-gate in a fault tree. In attack trees, the OR-gate represents a choice: the attacker
can choose which of the subgoals represented in the children to execute. In Fig. 16.2
the bank robber can choose whether to open the safe by unlocking it or by cutting
it open. In particular, the OR as a choice gate affects the probabilistic behavior.

Consider an intermediate goal A with independent subgoals B and C, connected via an OR-gate, i.e., $A = OR(B, C)$ and probabilities p_A, p_B, and p_C, respectively. In a fault tree, the probability for A is given by $p_A = p_B + p_C - p_B \cdot p_C$; see Sect. 10.3. However, attack trees assume that the attacker will maximize the chances on success, choosing the subgoal with the maximum probability, i.e., $p_A = \max(p_B, p_C)$.

Different Attributes. While quantitative analysis for fault trees is centered around probabilistic analysis, attack trees feature a multitude of different attributes. Apart from attack probabilities, the time of the shortest attack is relevant, its damage, and the skills that are required. In [12], a methodology is proposed that computes any of such algorithms, by observing that many attribute domains adhere to the axioms of a semi-ring, and that these semi-rings feature efficient algorithms via BDDs. Moreover, as often the case when multiple attributes come into play, these attributes can be conflicting. For example, one attack may be cheaper, while another takes less time. To make trade offs in such cases, Pareto analysis techniques have been developed.

16.1.3 Combining Attack Trees and Fault Trees

While fault trees focus on safety, and attack trees on security, it is widely acknowledged that effective risk management requires a joint analysis of safety and security risks [13]. An important reason is that measures for safety and security are often conflicting, that is, measures that increase safety may decrease security and vice versa. For example, locking the front door at night increases security, by preventing burglars to enter the house quickly. At the same time, it decreases safety, by disabling quick escapes in case of a fire. Such conflicts arise at large in industry: The Internet-of-Things offers excellent technology to monitor safety in production plants, but is notorious for being easy to hack, where the Vekada and 2010 Stuxnet attacks are infamous examples. Vice versa: Passwords secure people's medical data, but hinder quick access at emergencies, compromising patient safety. Therefore, safety and security must be analyzed in combination, otherwise measures taken will be counterproductive.

To accommodate this joint analysis, several combinations of attack trees and fault trees have been proposed, see [14] for an extensive overview. Fault trees/attack trees [15] start from the observation that a system can be corrupted if an attacker is able maliciously provoke a failure of one of the basic events. Thus, the basic events of a fault tree are further refined into an attack tree.

Attack-fault trees (AFTs) [16] merge dynamic attack trees and dynamic fault trees. Boolean-Driven Markov Processes (BDMPs) [17] extend attack trees and fault trees with triggers, modeling how a fault or attack triggers another one.

It is important to notice that none of these extensions take into account the different behaviors of the gates in fault trees and attack trees.

16.1.4 Component Fault Trees (CFTs)

Component fault trees (CFTs) equip fault trees with a modular structure [18], so that a large fault tree can be modeled and analyzed in terms of smaller components. Originating from fault tree analysis, CFTs enhance the traditional approach by modularizing the analysis process. This modularization allows for the examination of individual components within a system, facilitating a more detailed and scalable analysis of system reliability and safety [18]. In Fig. 16.3, possibilities for a safe trip to fail are analyzed with a CFT. Each subcomponent (*Road trip*, *Phone*, *Car*, *Engine*) is modeled and detailed separately, allowing for more granularity and scalability. CFTs are particularly useful in the context of systems engineering, where individual components' reliability can significantly impact the overall system's performance. By isolating components, engineers can identify potential failure points and assess their impact on the entire system. This approach not only aids in the design of more robust systems but also contributes to more efficient maintenance strategies by targeting specific components for improvement or monitoring [19]. Notably, CFTs have been used to attempt joint analysis of safety (no harm or risk due to accidental failures) and security (no harm or risk due to intentional attacks). For example, authors in [20] extend component fault trees with security aspects by introducing a new basic events type for security breaches, which are essentially basic attack steps.

16.1.5 Boolean-Driven Markov Models

Boolean-Driven Markov Processes (BDMPs) [21] extend fault trees via richer basic events and triggers.

Each basic event can be equipped with a Markov process (MP), representing the different modes a component can be in. Various templates provide standard MPs to model standard failure behavior. For example, the *failure in operation* MP contains two modes, *operational* and *failed*. One transitions from *operational* to *failed* with an exponential failure rate λ, and back to operational with a repair rate μ. The *failure on demand* MP models instantaneous failures. Moreover, users can define their own MPs as a stochastic Petri net [17].

The key construct of BDMPs, *triggers*, allows specifying functional dependencies or distinguishing subsystems in primary and cold stand-by. Triggers are represented by dotted red arrows, going from one gate to another gate. When the gate at the start of the arrow fails, then the trigger starts the subsystem modeled by the gate at the end of the arrow. Semantically, failure of the gate at the start of the arrow changes states in other MPs under the triggered gate. Figure 16.4 shows a BDMP with three subsystems. Failures of the primary subsystem are modeled by the gate OR_1 and its subtree. Gates OR_2 and OR_3 represent failures of the first and the second back-up system. The second back-up system starts only when both the primary and the first back-up system fails.

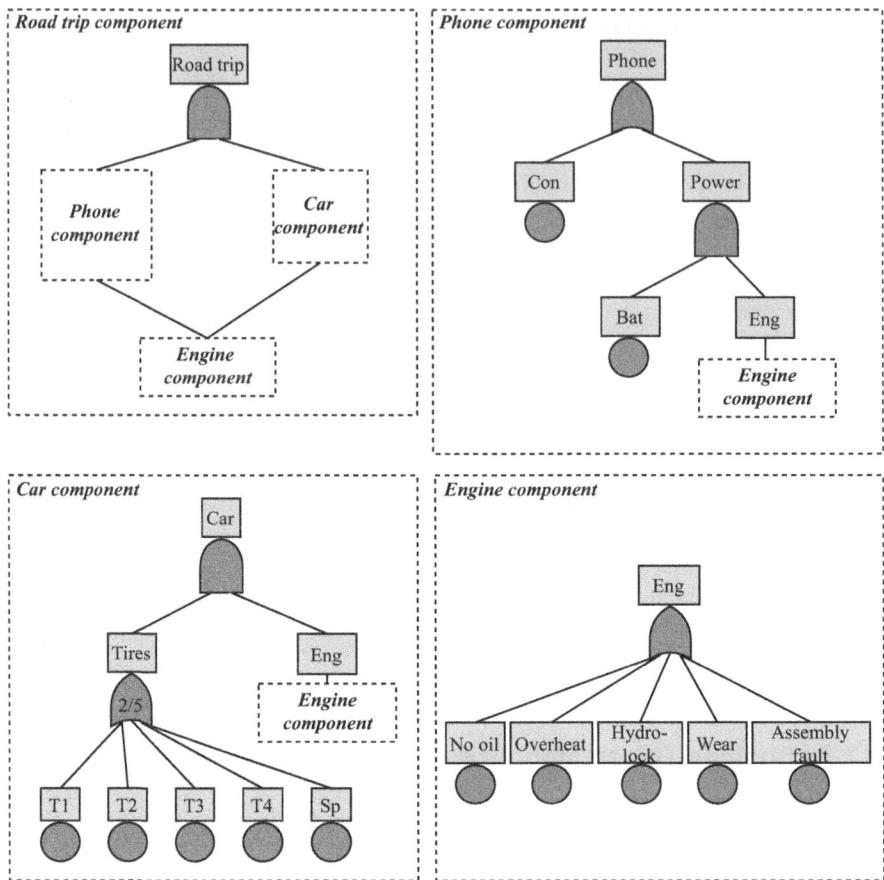

Fig. 16.3 CFT detailing failures on a road trip

System behaviors encoded by BDMPs belong to dynamic ones, same as for dynamic fault trees. In general, analysis of such systems must resort to Monte Carlo simulations. For certain subclasses of BDMPs, there are algorithms that first generate minimal cut sets of the underlying static fault tree and then quantify these minimal cut sets to approximate the top event probability [5–7].

16.1.6 Fault Trees with Complex Repairs

Standard fault trees provide only a limited model of repairable systems, where repairs are scheduled and performed entirely independently for each basic event. In practice, repairs of different components are often implicitly affected by each other: Repair staff and equipment are limited, so if multiple components have failed at the same time, operators may decide to repair the most critical components first, delaying

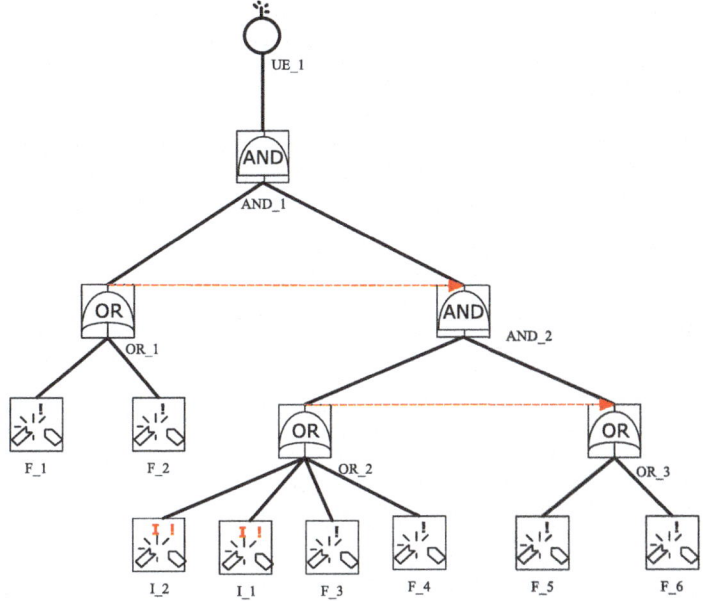

Fig. 16.4 Boolean-Driven Markov Process with two cascaded cold stand-by subsystems

repairs of the others. Extensions to dynamic fault trees have been developed [22] that allow specification of repair strategies, and analysis to determine how different strategies affect system reliability.

In addition to repairs in case of failures, many systems require routine maintenance to avoid failures. Furthermore, periodic inspections can be used to determine when preventive or corrective maintenance is necessary. A fault tree extension called "Fault maintenance trees" [23] has been developed to model such factors: Basic events are augmented with a degradation model, representing how component's condition goes down over time, eventually causing a failure. Periodic inspections are modeled to determine the current condition of a component, and repairs can be triggered by inspections, failures, or periodically, to return a component to its undegraded condition.

16.2   Extending Basic Events in Fault Trees

16.2.1   Fuzzy Fault Trees

In many practical settings, the failure probabilities of basic events are not known with a high precision. One approach to perform analysis in the face of such uncertainty is to use fault trees with fuzzy numbers [24].

Here, each failure probability is represented by a range of possible values, where the true probability is expected to be in this range. The possible ranges are described by *membership functions*, describing how strongly a given probability is contained in the range as described in Example 44.

When a fault tree is decorated with fuzzy failure probabilities, quantitative analysis can calculate similar metrics as for fault trees with normal, numeric, probabilities, except that the calculated metrics also have associated uncertainties. For an overview of analysis methods, and various variations of fuzzy fault trees, we refer the reader to [25].

Example 44 (Fuzzy probabilities)

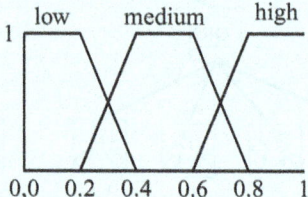

Fig. 16.5 Example fuzzy membership functions for the sets "low," "medium," and "high"

Suppose we try to estimate some failure probability that we are very uncertain about. To avoid needing to guess an exact number, we define categories "low," "medium," and "high." Figure 16.5 provides possible membership functions defining these categories. Here, probabilities below 0.2 are defined to be fully in the set "low," meaning that if the true probability is below 0.2, we are confident that we will estimate "low." The membership functions overlap to indicate uncertainty in our ratings. For example, the probability 0.7 has equal membership in "medium" and "high," meaning that if the true probability is 0.7, we are equally likely to estimate "medium" as "high."

16.2.2   Extended Fault Trees

More systems and components experience more complex failure behavior than just "failing," as various forms of failure and degraded operation can occur. Furthermore, such degraded operation in one component can affect the behavior of others. For example, an electric pump could fail to operate entirely, but it could also pump more slowly, or pump normally but draw excessive electrical current causing a fuse to eventually fail.

The formalism of Extended Fault Trees (EFT) [26] is an extension to cover such advanced failure behavior. It models basic events with multiple (mutually exclusive) states and transition rates between them. A textual language allows descriptions of

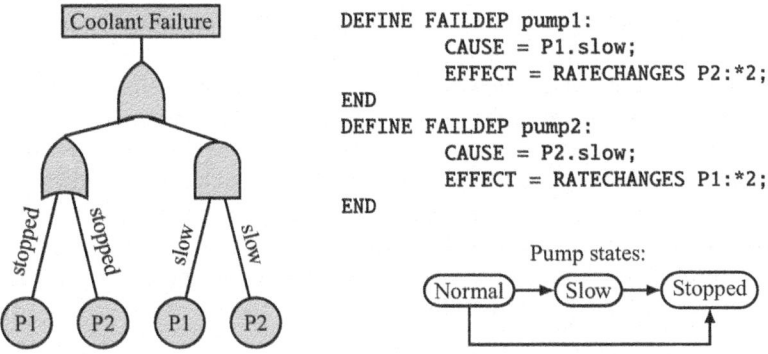

```
DEFINE FAILDEP pump1:
        CAUSE = P1.slow;
        EFFECT = RATECHANGES P2:*2;
END
DEFINE FAILDEP pump2:
        CAUSE = P2.slow;
        EFFECT = RATECHANGES P1:*2;
END
```

Fig. 16.6 Example of an extended FT. Pumps P1 and P2 have a normal state "Normal," and degraded modes "Slow" and "Stopped." If either pump is stopped, or if both pumps are slowed, the coolant system is considered failed. If either pump is slowed, the failure rates of the remaining pump are doubled

the dependencies between components. Figure 16.6 shows an example of such an extended fault tree.

Quantitative analysis of Extended fault trees is performed by translating the EFT to a Petri net [27], from which a range of quantitative metrics can be calculated.

16.3 Domain-Specific Fault Trees

16.3.1 Phased Fault Trees

Many systems are used in discrete "phases" with different failure behavior. For example, an airplane running out of fuel while taxiing on the ground is basically fine, while the same situation while flying above the ocean is often catastrophic.

Such systems can be modeled using standard fault trees, combining fault trees for phases into one fault tree using an OR-gate[28]. Because the phases will often share many components, techniques have been developed to speed up analysis in these cases [28,29].

16.3.2 Software Fault Trees

While fault trees have been widely deployed for dependability analysis of physical systems, they can also be used to analyze the reliability of software.

A fault tree for a software system is ultimately the same as a regular fault tree, the main difference lying in the method for constructing the tree: Because the behavior of a program is almost fully defined by its source code (and undefined situations, such as race conditions, are generally considered bugs in and of themselves), the potential causes of a failure condition in a simple program (fragment) can be derived algorithmically and described in a fault tree [30].

This automated approach, however, does not directly work for non-trivial programs. For complex program constructs, the fault tree quickly becomes impractically large. In fact, in the general case, the halting theorem means that even some modestly sized program exhibit behavior for which no fault tree can be automatically derived.

Software fault tree analysis remains an active area of research. While tools and techniques have been developed and applied to analyze critical software like fighter aircraft controls [31], there is currently no consensus on the best approach, and new approaches are still being proposed [32].

16.4 Related Reliability Models

16.4.1 Event Trees

While fault trees analyze the causes of system failures, event trees study potential consequences of a so-called *initiating event*, bringing the system out of a stable state. The system has typically safety barriers that shall mitigate the effect of the initiating event and bring the plant back to a stable, safe state. Event trees define accident sequences by an initial event followed by function events modeling failure or success of the safety barriers. Each accident sequence starts with the initiating event and decides for each safety barrier whether it failed or succeeded. Finally, each sequence is labeled by the set of consequences to which it leads.

Figure 16.7 shows an event tree with the initiating event LMFW and four function events. Lines in the graphical part of the tree denote sequences. If a line turns down below a function event then it means that this function (safety barrier) failed in the sequence. If the line continues straight then the function succeeded. The first and the third sequence in this event tree do not lead to a reactor core damage. Other sequences do.

Many models combine event trees and fault trees by using the latter to model failures of mitigating systems. In spite of a different graphical form and a different modeling purpose, event trees can be translated to fault trees. Probabilistic models of nuclear power plants contain event trees where some sequences lead to a reactor core damage as the consequence. By this we model the fact that the reactor core damage occurs when the accident evolves along one of these sequences. A software tool takes these sequences from the event trees together with the fault trees and transforms them into one, possibly very large, fault tree, called *master fault tree*. The top gate of this fault tree models the occurrence of the consequence.

Bow Ties Diagrams Another way to combine a fault tree and an event tree is using a "bowtie diagram." In such a diagram, a central (undesired) event is taken at the center of the diagram. A fault tree describing the causes of this event is drawn (rotated 90°) to the left of the undesired event, while a range of possible consequences and barriers limiting these consequences are shown in a event tree to the right.

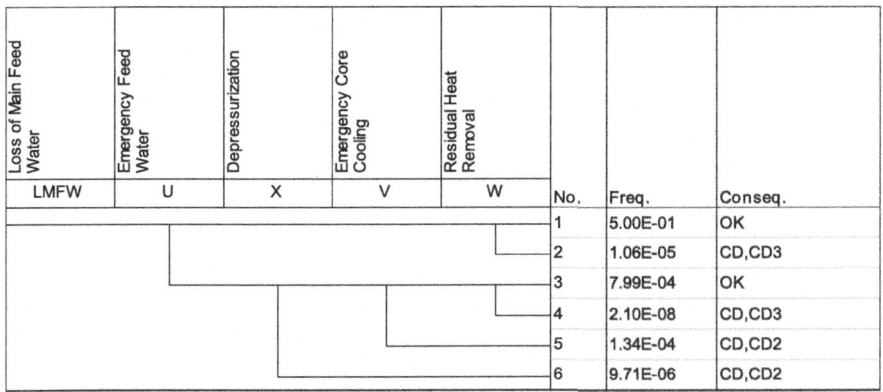

Fig. 16.7 An example event tree for accidents starting with a loss of main feedwater in a nuclear power plant

While quantitative analysis of bowtie diagrams is possible, it is common to use these diagrams for qualitative exploration, description, and communication of possible risks and mitigations [33].

References

1. Stamatelatos M, Vesely W, Dugan JB, Fragola J, Minarick J, Railsback J (2002) Fault tree handbook with aerospace applications. Office of safety and mission assurance NASA headquarters
2. Boudali H, Dugan JB (2005) A new Bayesian network approach to solve dynamic fault trees. In: Proceedings of the reliability and maintainability symposium (RAMS), IEEE, pp 451–456. https://doi.org/10.1109/RAMS.2005.1408404
3. Junges S, Guck D, Katoen JP, Stoelinga M (2016) Uncovering dynamic fault trees. In: Proceedings of the 46th annual IEEE/IFIP international conference on dependable systems and networks (DSN), IEEE, pp 299–310, https://doi.org/10.1109/DSN.2016.35
4. Singh C (1980) A cut set method for reliability evaluation of systems having s-dependent components. IEEE Trans Reliab R-29(5):372–375. https://doi.org/10.1109/TR.1980.5220886
5. Krčál J, Krčál P (2015) Scalable analysis of fault trees with dynamic features. In: Proceedings of the 45th annual IEEE/IFIP international conference on dependable systems and networks (DSN), IEEE, pp 89–100. https://doi.org/10.1109/DSN.2015.29
6. Bäckström O, Butkova Y, Hermanns H, Krčál J, Krčál P (2016) Effective static and dynamic fault tree analysis. In: Skavhaug A, Guiochet J, Bitsch F (eds) Proceedings of the international symposium on computer safety, reliability, and security (SAFECOMP), Springer, Lecture Notes on Computer Science, vol 9922, pp 266–280. https://doi.org/10.1007/978-3-319-45477-1_21
7. Bouissou M, Hernu O (2016) Boolean approximation for calculating the reliability of a very large repairable system with dependencies among components. In: Proceedings of the European safety and reliability conference (ESREL), CRC Press, Glasgow, UK, p 244

8. Walker MD (2009) Pandora: a logic for the qualitative analysis of temporal fault trees. PhD thesis, University of Hull

9. Schneier B (1999) Attack trees. Dr Dobb's Journal 24(12):21–29

10. Mauw S, Oostdijk M (2006) Foundations of Attack Trees. In: Won DH, Kim S (eds) International conference on information security and cryptology (ICISC), Springer, Lecture Notes on Computer Science, vol 3935, pp 186–198. https://doi.org/10.1007/11734727_17

11. Budde CE, Kolb C, Stoelinga M (2021) Attack trees vs. fault trees: Two sides of the same coin from different currencies. In: Abate A, Marin A (eds) Proceedings of the 18th International Conference on Quantitative Evaluation of Systems (QEST), Springer, Lecture Notes on Computer Science, vol 12846, pp 457–467, https://doi.org/10.1007/978-3-030-85172-9_24

12. Lopuhaä-Zwakenberg M, Budde CE, Stoelinga M (2023) Efficient and generic algorithms for quantitative attack tree analysis. IEEE Trans Dependable Secure Comput 20(5):4169–4187. https://doi.org/10.1109/TDSC.2022.3215752

13. Kriaa S, Piètre-Cambacédès L, Bouissou M, Halgand Y (2015) A survey of approaches combining safety and security for industrial control systems. Reliabil Eng System Safety 139:156–178. https://doi.org/10.1016/j.ress.2015.02.008

14. Nicoletti SM, Peppelman M, Kolb C, Stoelinga M (2023) Model-based joint analysis of safety and security: survey and identification of gaps. Comput Sci Rev 50. https://doi.org/10.1016/j.cosrev.2023.100597

15. Fovino IN, Masera M, De Cian A (2009) Integrating cyber attacks within fault trees. Reliabil Eng Syst Safety 94(9):1394–1402. https://doi.org/10.1016/j.ress.2009.02.020

16. Kumar R, Stoelinga M (2017) Quantitative security and safety analysis with attack-fault trees. In: Proceedings of the 18th IEEE international symposium on high-assurance systems engineering (HASE), pp 25–32, https://doi.org/10.1109/HASE.2017.12

17. Kriaa S, Bouissou M, Colin F, Halgand Y, Pietre-Cambacedes L (2014) Safety and security interactions modeling using the BDMP formalism: case study of a pipeline. In: Bondavalli A, Di Giandomenico F (eds) Proceedings of the 33rd international symposium on computer safety, reliability, and security (SAFECOMP), Springer, Lecture Notes in Computer Science, pp 326–341. https://doi.org/10.1007/978-3-319-10506-2_22

18. Kaiser B, Liggesmeyer P, Mäckel O (2003) A new component concept for fault trees. In: Proceedings of the 8th Australian workshop on safety critical systems and software (SCS'03), Australian computer Society, pp 37–46

19. Kaiser B, Schneider D, Adler R, Domis D, Möhrle F, Berres A, Zeller M, Höfig K, Rothfelder M (2018) Advances in component fault trees. In: Proceedings of the 28th European safety and reliability conference (ESREL). https://doi.org/10.1201/9781351174664-103

20. Steiner M, Liggesmeyer P (2013) Combination of safety and security analysis - finding security problems that threaten the safety of a system. In: Workshop DECS (ERCIM/EWICS workshop on dependable embedded and cyber-physical systems) of the 32nd international symposium on computer safety, reliability, and security (SAFECOMP)

21. Bouissou M (2002) Boolean logic Driven Markov Processes: a powerful new formalism for specifying and solving very large Markov models. In: Bonano EJ (ed) Proceedings of the 6th international conference on probabilistic safety assessment and management (PSAM), Elsevier, San Juan, Puerto Rico, USA

22. Guck D, Katoen JP, Stoelinga MIA, Luiten T, Romijn J (2014) Smart railroad maintenance engineering with stochastic model checking. In: Proceedings of the 2nd international conference on railway technology: research, development and maintenance (Railways), Civil-Comp Press, Stirlingshire, UK, Civil-Comp Proceedings, vol 104. https://doi.org/10.4203/ccp.104.299

23. Ruijters E, Guck D, Drolenga P, Stoelinga M (2016) Fault maintenance trees: reliability centered maintenance via statistical model checking. In: Proceedings of the IEEE 62nd annual reliability and maintainability symposium (RAMS), IEEE. https://doi.org/10.1109/RAMS.2016.7447986

24. Tanaka H, Fan L, Lai F, Toguchi K (1983) Fault-tree analysis by fuzzy probability. IEEE Trans Reliab 32(5):453–457. https://doi.org/10.1109/TR.1983.5221727

25. Mahmood YA, Ahmadi A, Verma AK, Srividya A, Kumar U (2013) Fuzzy fault tree analysis: a review of concept and application. Int J Syst Assur Eng Manag. pp 19–32. https://doi.org/10.1007/s13198-013-0145-x

26. Buchacker K (2000) Modeling with extended fault trees. In: Proceedings of the 5th IEEE international symposium on high-assurance systems engineering (HASE), pp 238–246. https://doi.org/10.1109/HASE.2000.895468

27. Buchacker K (1999) Combining fault trees and Petri nets to model safety-critical systems. In: Proceedings of the high performance computing symposium (HPC), The society for computer simulation international, pp 439–444

28. Meshkat L, Xing L, Donohue SK, Ou Y (2003) An overview of the phase-modular fault tree approach to phased mission system analysis. 2014/7138

29. Vaurio J (2001) Fault tree analysis of phased mission systems with repairable and non-repairable components. Reliab Eng Syst Safety 74(2):169–180. https://doi.org/10.1016/S0951-8320(01)00075-8

30. Leveson NG, Harvey PR (1983) Analyzing software safety. IEEE Trans Softw Eng SE-9(5):569–579. https://doi.org/10.1109/TSE.1983.235116

31. Weber W, Tondok H, Bachmayer M (2003) Enhancing software safety by fault trees: experiences from an application to flight critical SW. In: Proceedings of the international symposium on computer safety, reliability, and security (SAFECOMP), Springer, Lecture Notes on Computer Science, vol 2788, pp 289–302, https://doi.org/10.1007/978-3-540-39878-3_23

32. Takahashi M, Anang Y, Watanabe Y (2020) A proposal of fault tree analysis for embedded control software. Information 11(9):402. https://doi.org/10.3390/info11090402

33. de Ruijter A, Guldenmund F (2016) The bowtie method: a review. Safety Sci 88:211–218

Index

© The Editor(s) (if applicable) and The Author(s), under exclusive license to Springer
Nature Switzerland AG 2026
M. Stoelinga et al., *Concise Guide to Fault Tree Analysis*, Computer Science Foundations
and Applied Logic, https://doi.org/10.1007/978-3-031-78287-9

The manufacturer's authorised representative in the EU is Springer
Nature Customer Service Centre GmbH, Europaplatz 3, 69115 Heidelberg,
Germany. If you have any concerns regarding our products, please
contact ProductSafety@springernature.com

Printed and bound by CPI Group (UK) Ltd, Croydon, CR0 4YY
23/04/2026
02095585-0002